T0269288

SpringerBriefs in Mathematical Physics

Volume 10

More information about this series at http://www.springer.com/series/11953

Akira Furusawa

Quantum States of Light

 Springer

Akira Furusawa
Department of Applied Physics
The University of Tokyo
Tokyo
Japan

ISSN 2197-1757   ISSN 2197-1765   (electronic)
SpringerBriefs in Mathematical Physics
ISBN 978-4-431-55958-0   ISBN 978-4-431-55960-3   (eBook)
DOI 10.1007/978-4-431-55960-3

Library of Congress Control Number: 2015957783

Printed on acid-free paper

This Springer imprint is published by SpringerNature
The registered company is Springer Japan KK

Preface

We learn about the properties of quantized optical fields in quantum optics. Although this may sound old and traditional, it is not. In reality, we assumed optical fields as classical fields until very recently. We did not have to quantize the field because our light source was only a laser and whose state, a coherent state, can be regarded as a classical field.

We have to use quantum optics nowadays, of course. It is because squeezed light is easily created these days and we have to handle it. Squeezed light is a pure quantum mechanical state, which cannot be described without quantum optics. In that sense, the "phase transition" occurred when Slusher et al. created the squeezed light for the very first time in 1985. After the "phase transition," various "pure" quantum states were created, which include superposition of a vacuum and a single-photon state, a Schrödinger's cat state, and so on.

In this book, we explain the definition and the way to create these "modern" quantum states of light. For that purpose we use many figures to visualize the quantum states to help the readers' understanding, because the quantum states sometimes look very counterintuitive when one only looks at equations.

Tokyo   Akira Furusawa

Acknowledgments

All the experimental results presented in this book come from experiments carried out by the members of Furusawa group at the Department of Applied Physics of The University of Tokyo, including Nobuyuki Takei, Hidehiro Yonezawa, Yuishi Takeno, Jun-ichi Yoshikawa, Noriyuki Lee, Mitsuyoshi Yukawa, Yoshichika Miwa, Kazunori Miyata, and Maria Fuwa. The author would like to thank all members of Furusawa group. The author also acknowledges Ilian Häggmark for the language review.

This book was originally written in Japanese and published by Uchida Roukakuho in 2013.

Tokyo   Akira Furusawa

Contents

Chapter 1
Quantum States of Light

1.1 Quantization of Optical Fields

In this section, we present an intuitive description of quantum optics.

According to the quantum field theory a vector potential operator of optical fields $\hat{\mathcal{A}}(r, t)$ can be described as

$$\hat{\mathcal{A}}(r, t) = \mathcal{A}_0 e^{i(k \cdot r - \omega t)} \hat{a} + \mathcal{A}_0^* e^{-i(k \cdot r - \omega t)} \hat{a}^\dagger. \tag{1.1}$$

Here the spatial mode is a plane wave which propagates in the direction of the wave vector k, $\mathcal{A}_0$ denotes a complex vector potential orthogonal to the wave vector, ω is the angular frequency of the optical field, $\hat{a}$ and $\hat{a}^\dagger$ are the annihilation and creation operators, respectively, and $\hat{n} = \hat{a}^\dagger \hat{a}$ is the number operator.

$\hat{a}$ and $\hat{a}^\dagger$ act on eigenstates of the number operator (Fock states) $|n\rangle$ $(\hat{n}|n\rangle = n|n\rangle)$ as

$$\begin{aligned}
\hat{a}|n\rangle &= \sqrt{n}|n - 1\rangle, \\
\hat{a}^\dagger|n\rangle &= \sqrt{n + 1}|n + 1\rangle, \\
\hat{n}|n\rangle &= n|n\rangle.
\end{aligned} \tag{1.2}$$

We can create an electrical-field operator $\hat{\mathcal{E}}(r, t)$ and a magnetic-flux density operator $\hat{\mathcal{B}}(r, t)$ of the optical field by using the following equations and Eq. (1.1). Namely, by using

$$\begin{aligned}
\mathcal{E}(r, t) &= -\frac{\partial \mathcal{A}(r, t)}{\partial t}, \\
\mathcal{B}(r, t) &= \nabla \times \mathcal{A}(r, t)
\end{aligned} \tag{1.3}$$

and Eq. (1.1), we can get

© The Author(s) 2015
A. Furusawa, *Quantum States of Light*, SpringerBriefs
in Mathematical Physics, DOI 10.1007/978-4-431-55960-3_1

$$\hat{\mathcal{E}}(r, t) = i\omega(\mathcal{A}_0 e^{i(k \cdot r - \omega t)} \hat{a} - \mathcal{A}_0^* e^{-i(k \cdot r - \omega t)} \hat{a}^\dagger), \tag{1.4}$$

$$\hat{\mathcal{B}}(r, t) = ik \times (\mathcal{A}_0 e^{i(k \cdot r - \omega t)} \hat{a} - \mathcal{A}_0^* e^{-i(k \cdot r - \omega t)} \hat{a}^\dagger). \tag{1.5}$$

Moreover, from $k \cdot \mathcal{A}_0 = 0$ and $|\mathcal{A}_0| = \mathcal{A}_0$, we can get the Hamiltonian $\hat{H}$ which corresponds to the field energy as

$$
\begin{aligned}
\hat{H} &= \int \left(\frac{1}{2} \epsilon_0 \hat{\mathcal{E}}(r, t) \cdot \hat{\mathcal{E}}(r, t) + \frac{1}{2\mu_0} \hat{\mathcal{B}}(r, t) \cdot \hat{\mathcal{B}}(r, t) \right) dr \\
&= \frac{1}{2} \left(\epsilon_0 \omega^2 + \frac{1}{\mu_0} |k|^2 \right) \int \mathcal{A}_0^2 dr \, (\hat{a}\hat{a}^\dagger + \hat{a}^\dagger \hat{a}) \\
&= \frac{\hbar\omega}{2} (\hat{a}\hat{a}^\dagger + \hat{a}^\dagger \hat{a}) \\
&= \hbar\omega \left(\hat{n} + \frac{1}{2} \right),
\end{aligned}
\tag{1.6}
$$

where ϵ_0 is the permittivity of vacuum, μ_0 is the magnetic permeability of vacuum, $\int \mathcal{A}_0^2 dr = \hbar/2\epsilon_0\omega$, and $\int e^{\pm 2ik \cdot r} dr = 0$.

Now let the optical field operators evolve in time. When the Hamiltonian does not change in time, the Heisenberg equation of motion of an operator $\hat{A}(t)$ becomes

$$i\hbar \frac{d\hat{A}(t)}{dt} = [\hat{A}(t), \hat{H}]. \tag{1.7}$$

By using this equation, we can get

$$\hat{A}(t) = e^{i\frac{\hat{H}}{\hbar}t} \hat{A}(0) e^{-i\frac{\hat{H}}{\hbar}t}. \tag{1.8}$$

So the time evolution of an electrical-field operator $\hat{\mathcal{E}}(r, t)$ of an optical field should obey

$$\hat{\mathcal{E}}(r, t) = e^{i\frac{\hat{H}}{\hbar}t} \hat{\mathcal{E}}(r, 0) e^{-i\frac{\hat{H}}{\hbar}t}. \tag{1.9}$$

By using Eqs. (1.6) and (1.9) we can check Eq. (1.4) from the view point of time evolution of operators. Note that we used the following equation here:

$$e^{i\frac{\hbar\omega(\hat{n}+1/2)}{\hbar}t} \hat{a} e^{-i\frac{\hbar\omega(\hat{n}+1/2)}{\hbar}t} = \hat{a} e^{-i\omega t}. \tag{1.10}$$

Similarly we can check Eq. (1.5) for the magnetic-flux density operator $\hat{\mathcal{B}}(r, t)$ of an optical field from the view point of time evolution with Eqs. (1.6) and (1.9). Although everything is "peaceful and quiet" so far, we do more in quantum optics.

In quantum optics we think that an annihilation operator $\hat{a}$ evolves according to Eq. (1.10). Namely we set $\hat{a}(t) = \hat{a} e^{-i\omega t}$. It is a misunderstanding in some sense, because an annihilation operator is a field operator and should not evolve in time. However, if we set it like this, it becomes very convenient. *So in quantum optics we treat an annihilation operator as a complex amplitude of an optical field.* Of course, we know that it is a "dialect" which only works in quantum optics.

We will explain the "dialect" in more detail here. First of all, we define Hermitian operators $\hat{x}$ and $\hat{p}$, which are called quadrature operators, as follows:

$$\hat{x} = \sqrt{\frac{\hbar}{2\omega}}(\hat{a} + \hat{a}^{\dagger}),$$
$$\hat{p} = \frac{1}{i}\sqrt{\frac{\hbar\omega}{2}}(\hat{a} - \hat{a}^{\dagger}). \tag{1.11}$$

It follows from $[\hat{a}, \hat{a}^{\dagger}] = 1$ that $[\hat{x}, \hat{p}] = i\hbar$ and $\hat{x}$ and $\hat{p}$ satisfy the same commutation relationship as the one for position and momentum operators, which means that the quadrature operators are conjugated variables like position and momentum. $\hat{x}$ and $\hat{p}$ are thus sometimes called generalized position and momentum. Of course, they are totally different from position and momentum, though.

Next we rearrange Eq. (1.11) and get the following equations:

$$\hat{a} = \sqrt{\frac{\omega}{2\hbar}}\hat{x} + i\sqrt{\frac{1}{2\hbar\omega}}\hat{p},$$
$$\hat{a}^{\dagger} = \sqrt{\frac{\omega}{2\hbar}}\hat{x} - i\sqrt{\frac{1}{2\hbar\omega}}\hat{p}. \tag{1.12}$$

By using these equations we can rewrite the electrical-field operator of an optical field $\hat{\mathcal{E}}(\boldsymbol{r}, t)$ (Eq. (1.4)) with $\hat{x}$ and $\hat{p}$ and get the following equation:

$$\hat{\mathcal{E}}(\boldsymbol{r}, t) = -2\omega\mathcal{A}_0\boldsymbol{e}\left[\sqrt{\frac{\omega}{2\hbar}}\hat{x}\sin(\boldsymbol{k} \cdot \boldsymbol{r} - \omega t) + \sqrt{\frac{1}{2\hbar\omega}}\hat{p}\cos(\boldsymbol{k} \cdot \boldsymbol{r} - \omega t)\right], \quad (1.13)$$

where we assume that $\boldsymbol{e} = \mathcal{A}_0/\mathcal{A}_0$ and $\mathcal{A}_0$ is a real number. We do not loose the generality by this assumption, because there is no absolute phase of an optical field.

We make a more "rough" treatment in quantum optics. That is, we select some proper unit and let $\omega = 1$, $\hbar = 1/2$, and $2\mathcal{A}_0 = 1$. As the result Eq. (1.13) becomes

$$\hat{\mathcal{E}}(\boldsymbol{r}, t) = -\boldsymbol{e}[\hat{x}\sin(\boldsymbol{k} \cdot \boldsymbol{r} - \omega t) + \hat{p}\cos(\boldsymbol{k} \cdot \boldsymbol{r} - \omega t)]. \tag{1.14}$$

Here we keep ω in sine and cosine to stress the temporal dependence.

With $\omega = 1$, $\hbar = 1/2$, and $2\mathcal{A}_0 = 1$, Eq. (1.12) becomes

$$\hat{a} = \hat{x} + i\hat{p},$$
$$\hat{a}^{\dagger} = \hat{x} - i\hat{p}, \tag{1.15}$$

and, of course, $[\hat{x}, \hat{p}] = i/2$.

Comparing Eq. (1.14) with Eq. (1.15) we can say that we might be able to treat an annihilation operator $\hat{a}$ as a complex amplitude of an optical electric-field as we mentioned before. It means that $\hat{x}$ and $\hat{p}$ look like the amplitudes of sine and cosine components of the complex amplitude $\hat{a}$, respectively. Here $\hat{x}$ and $\hat{p}$ are called quadrature amplitude operators. It is of course a "distorted interpretation", but it is a "de facto standard" nowadays. However, it is a big benefit that we can intuitively understand quantum-optics experiments. In any case, we should not forget the importance of Eqs. (1.4) and (1.10). In other words, we can do anything we like in quantum optics if we do not forget it.

Note that there is another way to simplify Eq. (1.12), where we take $\hbar = 1$ instead of $\hbar = 1/2$. Then we get

$$\hat{a} = \frac{\hat{x} + i\hat{p}}{\sqrt{2}},$$
$$\hat{a}^\dagger = \frac{\hat{x} - i\hat{p}}{\sqrt{2}},$$

(1.16)

and $[\hat{x}, \hat{p}] = i$. $\hbar = 1$ and $\hbar = 1/2$ are both frequently used and it is therefore sometimes really confusing. We take $\hbar = 1/2$ in this book except for a few cases where we explicitly declare to take $\hbar = 1$. This is because the author prefers $\hbar = 1/2$. There is of course no essential difference between them.

There is one more thing. That is the uncertain relationship between $\hat{x}$ and $\hat{p}$. General operators $\hat{A}$ and $\hat{B}$ satisfy the following inequality for an arbitrary state $|\psi\rangle$:

$$\sqrt{\langle(\Delta\hat{A})^2\rangle\langle(\Delta\hat{B})^2\rangle} \geq \frac{1}{2}\left|\langle[\hat{A}, \hat{B}]\rangle\right|,$$

(1.17)

where,

$$\langle(\Delta\hat{A})^2\rangle \equiv \langle\psi|\hat{A}^2|\psi\rangle - \langle\psi|\hat{A}|\psi\rangle^2.$$

(1.18)

When $\hbar = 1/2$, the commutation relationship between $\hat{x}$ and $\hat{p}$ is $[\hat{x}, \hat{p}] = i/2$. Then Eq. (1.17) for $\hat{x}$ and $\hat{p}$ becomes

$$\sqrt{\langle(\Delta\hat{x})^2\rangle\langle(\Delta\hat{p})^2\rangle} \geq \frac{1}{4}.$$

(1.19)

This is the uncertainty relationship between $\hat{x}$ and $\hat{p}$.

1.2 Coherent States

We explained the "de facto standard" of quantum optics in the previous section. Now we will describe various quantum states of light by using this standard. As a first example we describe coherent states, which correspond to a state of laser light. The formalization of quantum optics was made by Glauber just after the invention of the laser and it was actually for the quantum state of laser light. In that sense coherent states should be called Glauber states.

The coherent state $|\alpha\rangle$ is defined as an eigenstate of an annihilation operator $\hat{a}$ as

$$\hat{a}|\alpha\rangle = \alpha|\alpha\rangle, \tag{1.20}$$

where α corresponds to the complex amplitude of the optical electric field. As the readers might notice, this is the origin of the "de facto standard" of quantum optics, namely this is the reason why we can treat an annihilation operator as the complex amplitude of the optical electric field. Of course, we know that an annihilation operator $\hat{a}$ is not a Hermitian operator, which means it is not an observable.

From the definition of coherent states (Eq. (1.20)),

$$\langle\alpha|\hat{a}^{\dagger} = \langle\alpha|\alpha^{*}, \tag{1.21}$$

and with $[\hat{a}, \hat{a}^{\dagger}] = 1$ and Eq. (1.15), we can derive the following equations:

$$\begin{aligned} \langle\alpha|\hat{x}|\alpha\rangle &= \langle\alpha|\frac{\hat{a} + \hat{a}^{\dagger}}{2}|\alpha\rangle \\ &= \frac{\alpha + \alpha^{*}}{2} \\ &= \Re\{\alpha\}, \end{aligned} \tag{1.22}$$

$$\begin{aligned} \langle\alpha|\hat{p}|\alpha\rangle &= \langle\alpha|\frac{\hat{a} - \hat{a}^{\dagger}}{2i}|\alpha\rangle \\ &= \frac{\alpha - \alpha^{*}}{2i} \\ &= \Im\{\alpha\}, \end{aligned} \tag{1.23}$$

$$\begin{aligned} \langle\alpha|\hat{x}^{2}|\alpha\rangle &= \langle\alpha|\left(\frac{\hat{a} + \hat{a}^{\dagger}}{2}\right)^{2}|\alpha\rangle \\ &= \Re\{\alpha\}^{2} + \frac{1}{4}, \end{aligned} \tag{1.24}$$

$$\begin{aligned} \langle\alpha|\hat{p}^{2}|\alpha\rangle &= \langle\alpha|\left(\frac{\hat{a} - \hat{a}^{\dagger}}{2i}\right)^{2}|\alpha\rangle \\ &= \Im\{\alpha\}^{2} + \frac{1}{4}, \end{aligned} \tag{1.25}$$

$$\langle (\Delta \hat{x})^2 \rangle \equiv \langle \alpha | \hat{x}^2 | \alpha \rangle - \langle \alpha | \hat{x} | \alpha \rangle^2$$
$$= \Re\{\alpha\}^2 + \frac{1}{4} - \Re\{\alpha\}^2$$
$$= \frac{1}{4}, \qquad\qquad (1.26)$$

$$\langle (\Delta \hat{p})^2 \rangle \equiv \langle \alpha | \hat{p}^2 | \alpha \rangle - \langle \alpha | \hat{p} | \alpha \rangle^2$$
$$= \Im\{\alpha\}^2 + \frac{1}{4} - \Im\{\alpha\}^2$$
$$= \frac{1}{4}. \qquad\qquad (1.27)$$

From above equations we can say the following things.

From the "de facto standard" of quantum optics we can think of $\hat{x}$ and $\hat{p}$ as amplitudes of the sine and cosine components of the optical electric field, respectively, as in Eq. (1.14). So the mean values correspond to the ones of real and imaginary components of the complex amplitude. Moreover the standard deviations, i.e., square root of the variances $\langle (\Delta \hat{x})^2 \rangle$ and $\langle (\Delta \hat{p})^2 \rangle$ both become 1/4 in any case. It means that the standard deviations are always 1/4 irrespective of the complex amplitude. Therefore a coherent state in phase space is described as shown in Fig. 1.1. Note that a coherent state is a minimum uncertainty state which corresponds to the case where the equality holds in Eq. (1.19).

Fig. 1.1 A coherent state in phase space

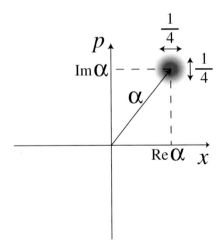

Next we show that a coherent state of light can be described by a superposition of the photon-number (Fock) states $|n\rangle$. It is possible because the photon-number states $|n\rangle$ are an orthonormal basis of quantum states of light. Namely,[1]

$$\langle l|m\rangle = \delta_{lm}, \tag{1.28}$$

and an arbitrary quantum state $|\psi\rangle$ can be described as

$$|\psi\rangle = \sum_{n=0}^{\infty} c_n |n\rangle, \tag{1.29}$$

where c_n is a complex number.

By using above equation a coherent state $|\alpha\rangle$ can be described as a superposition of photon-number states $|n\rangle$ with

$$|\alpha\rangle = \sum_{n=0}^{\infty} w_n |n\rangle. \tag{1.30}$$

From the definition of a coherent state (Eq. (1.20)) follows that

$$\hat{a}|\alpha\rangle = \alpha|\alpha\rangle = \sum_{n=0}^{\infty} \alpha w_n |n\rangle. \tag{1.31}$$

From Eq. (1.2),

$$\hat{a}|\alpha\rangle = \sum_{n=0}^{\infty} w_n \sqrt{n}|n-1\rangle. \tag{1.32}$$

From Eqs. (1.31) and (1.32) and taking into account that $|n\rangle$ is an orthonormal basis,

$$w_{n+1} = \alpha w_n. \tag{1.33}$$

Therefore,

$$\begin{aligned} w_n &= \frac{\alpha}{\sqrt{n}} w_{n-1} \\ &= \frac{\alpha}{\sqrt{n}} \frac{\alpha}{\sqrt{n-1}} \cdots \frac{\alpha}{1} w_0 \\ &= \frac{\alpha^n}{\sqrt{n!}} w_0. \end{aligned} \tag{1.34}$$

_____

[1] δ_{lm} is called Kronecker delta and it is 1 when $l = m$ and 0 when $l \neq m$.

Since $|\alpha\rangle$ has to be normalized, we can get the following equation by using Eqs. (1.31) and (1.34):

$$
\begin{aligned}
1 &= \langle\alpha|\alpha\rangle \\
&= \sum_{n=0}^{\infty} |w_n|^2 \\
&= \sum_{n=0}^{\infty} \frac{|\alpha|^{2n}}{\sqrt{n!}} |w_0|^2 \\
&= e^{|\alpha|^2} |w_0|^2.
\end{aligned}
\tag{1.35}
$$

Hence,

$$
|w_0|^2 = e^{-|\alpha|^2}.
\tag{1.36}
$$

In general, w_0 should satisfy

$$
w_0 = e^{-\frac{|\alpha|^2}{2} - i\phi}.
\tag{1.37}
$$

In the case of light we do not have to worry about the initial phase and we can set $\phi = 0$. Finally we get

$$
w_0 = e^{-\frac{|\alpha|^2}{2}}.
\tag{1.38}
$$

From above derivation we can get the following equation for a coherent state $|\alpha\rangle$:

$$
|\alpha\rangle = e^{-\frac{|\alpha|^2}{2}} \sum_{n=0}^{\infty} \frac{\alpha^n}{\sqrt{n!}} |n\rangle.
\tag{1.39}
$$

Since the probability distribution of photon number can be determined by the square of the coefficients of $|n\rangle$, we can get the following function of n for it:

$$
\frac{\left(|\alpha|^2\right)^n}{n!} e^{-|\alpha|^2}.
\tag{1.40}
$$

We know that it is a Poisson distribution with an averaged value of $|\alpha|^2$.

Let's think about the time evolution of a coherent state with a Hamiltonian $\hbar\omega(\hat{n} + 1/2)$ (Eq. (1.6)) and the following Schrödinger equation with a time-invariant Hamiltonian:

$$
|\psi(t)\rangle = e^{-i\frac{\hat{H}}{\hbar}t}|\psi(0)\rangle.
\tag{1.41}
$$

We set the initial state as $|\alpha\rangle$ and the final state as $|\alpha(t)\rangle$.

$$|\alpha(t)\rangle = e^{-i\frac{\hat{H}}{\hbar}t}|\alpha\rangle$$

$$= e^{-i\frac{\omega}{2}t}e^{-i\hat{n}\omega t}|\alpha\rangle$$

$$= e^{-i\frac{\omega}{2}t}\sum_{m=0}^{\infty}\frac{(-i\hat{n}\omega t)^m}{m!} \cdot e^{-\frac{|\alpha|^2}{2}}\sum_{n=0}^{\infty}\frac{\alpha^n}{\sqrt{n!}}|n\rangle$$

$$= e^{-i\frac{\omega}{2}t}\sum_{m=0}^{\infty}\frac{(-i\omega t)^m}{m!} \cdot e^{-\frac{|\alpha|^2}{2}}\sum_{n=0}^{\infty}\frac{\alpha^n}{\sqrt{n!}}\hat{n}^m|n\rangle$$

$$= e^{-i\frac{\omega}{2}t}\sum_{m=0}^{\infty}\frac{(-i\omega t)^m}{m!} \cdot e^{-\frac{|\alpha|^2}{2}}\sum_{n=0}^{\infty}\frac{\alpha^n}{\sqrt{n!}}n^m|n\rangle$$

$$= e^{-i\frac{\omega}{2}t}\sum_{m=0}^{\infty}\frac{(-in\omega t)^m}{m!} \cdot e^{-\frac{|\alpha|^2}{2}}\sum_{n=0}^{\infty}\frac{\alpha^n}{\sqrt{n!}}|n\rangle$$

$$= e^{-i\frac{\omega}{2}t}e^{-in\omega t} \cdot e^{-\frac{|\alpha|^2}{2}}\sum_{n=0}^{\infty}\frac{\alpha^n}{\sqrt{n!}}|n\rangle$$

$$= e^{-i\frac{\omega}{2}t} \cdot e^{-\frac{|\alpha|^2}{2}}\sum_{n=0}^{\infty}\frac{(\alpha e^{-i\omega t})^n}{\sqrt{n!}}|n\rangle$$

$$= e^{-i\frac{\omega}{2}t}|\alpha e^{-i\omega t}\rangle. \tag{1.42}$$

Since the phase factor for the overall state has no meaning here, we can say that the final state of the time evolution of a coherent state $|\alpha\rangle$ with a Hamiltonian $\hbar\omega(\hat{n} + 1/2)$ is $|\alpha e^{-i\omega t}\rangle$. Namely, the initial coherent state with a complex amplitude of α becomes the one with a complex amplitude of $\alpha e^{-i\omega t}$ after time t (Fig. 1.3). As will be explained later, although a quantum state of light changes through the time evolution in general, a coherent state keeps to a coherent state, and it only changes in phase as shown in Fig. 1.2. Moreover, the time evolution of a coherent state can also be

Fig. 1.2 Time evolution of a coherent state in phase space

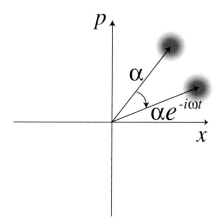

Fig. 1.3 Time dependence
of the real or the imaginary
part of the optical electric
field of a coherent state

plotted as a time dependence of the real or the imaginary part of the optical electric
field as shown in Fig. 1.3. This is just a wave!

Here we think about time dependence of general states $|\psi\rangle = \sum_{n=0}^{\infty} c_n|n\rangle$ (Eq. (1.29))
for the sake of showing the specialness of coherent states which keep the state through
the time evolution. Namely, we can time evolve $|\psi\rangle = \sum_{n=0}^{\infty} c_n|n\rangle$ with Eq. (1.41) as
follows:

$$|\psi(t)\rangle = e^{-i\frac{\hat{H}}{\hbar}t}|\psi(0)\rangle$$

$$= e^{-i\frac{\omega}{2}t} e^{-i\hat{n}\omega t} \sum_{n=0}^{\infty} c_n|n\rangle$$

$$= e^{-i\frac{\omega}{2}t} \sum_{n=0}^{\infty} c_n e^{-in\omega t}|n\rangle. \tag{1.43}$$

This result shows that the phase frequency is proportional to photon number n, where
we neglect the overall phase factor $e^{-i\frac{\omega}{2}t}$. So the phase does not change when $n = 0$
and it does change at the frequency rate of ω when $n = 1$. So far everything seems
to be trivial, however, the phase frequency becomes 2ω when $n = 2$ and it becomes
$n\omega$ when the photon number is n. Although coherent states keep the state through
the time evolution as pointed out above (Fig. 1.2), it is not true in general because the
phase frequency depends on the photon number and Eq. (1.43) cannot be simplified
like Eq. (1.42).

1.3 Balanced Homodyne Measurement

We can check the time dependence of the coherent state shown in Fig. 1.3 by experiments. It is called balanced homodyne measurement. Figure 1.4 shows the schematic
of balanced homodyne measurement. Here the measured light beam (spatial mode)
is $\hat{a}_1$ and the light beam $\hat{a}_2$ is called a local oscillator which is a large-amplitude
coherent light beam. It is convenient to use the "de facto standard" of quantum

Fig. 1.4 Balanced
homodyne measurement.
Here the measured light
beam (spatial mode) is $\hat{a}_1$
and the light beam $\hat{a}_2$ is
called a local oscillator
which is a large-amplitude
coherent light beam

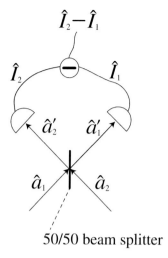

optics; $\hat{a}$ can be treated as the quantum complex amplitude, in order to understand
the mechanism of balanced homodyne measurement. On top of that, we should understand the unitary transformation of beam splitters for quantum complex amplitudes.
So we will explain beam splitters as quantum devices in the next section.

1.3.1 Beam Splitters

A beam splitter is an optical device which has two input light beams $\hat{a}_1$ and $\hat{a}_2$ and two
output light beams $\hat{a}_1'$ and $\hat{a}_2'$. The input-output relation, i.e., the relationship between
$(\hat{a}_1, \hat{a}_2)$ and $(\hat{a}_1', \hat{a}_2')$, can be described by a 2×2 matrix $\underline{B}$ as follows (Fig. 1.5):

$$\begin{pmatrix} \hat{a}_1' \\ \hat{a}_2' \end{pmatrix} = \underline{B} \begin{pmatrix} \hat{a}_1 \\ \hat{a}_2 \end{pmatrix} = \begin{pmatrix} B_{11} & B_{12} \\ B_{21} & B_{22} \end{pmatrix} \begin{pmatrix} \hat{a}_1 \\ \hat{a}_2 \end{pmatrix}. \tag{1.44}$$

Fig. 1.5 A beam splitter

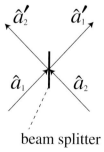

B cannot be an arbitrary matrix but has a constraint. That is energy conservation, i.e., the total photon number should be kept during the beam-splitter transformation:

$$\hat{a}_1^\dagger \hat{a}_1 + \hat{a}_2^\dagger \hat{a}_2 = \hat{a}_1'^\dagger \hat{a}_1' + \hat{a}_2'^\dagger \hat{a}_2'. \tag{1.45}$$

From Eq. (1.45), B should satisfy the following conditions:

$$|B_{11}|^2 + |B_{21}|^2 = |B_{12}|^2 + |B_{22}|^2, \tag{1.46}$$

$$B_{11}^* B_{12} + B_{21}^* B_{22} = 0. \tag{1.47}$$

These conditions are actually equivalent to the ones for the unitarity of matrix B, i.e., $\underline{B}^\dagger \underline{B} = E$.

An arbitrary 2×2 unitary matrix can be described with real numbers Λ, Ψ, Θ, and Φ as follows:

$$\underline{B} = e^{i\Lambda/2} \begin{pmatrix} e^{i\Psi/2} & 0 \\ 0 & e^{-i\Psi/2} \end{pmatrix} \begin{pmatrix} \cos(\Theta/2) & \sin(\Theta/2) \\ -\sin(\Theta/2) & \cos(\Theta/2) \end{pmatrix} \begin{pmatrix} e^{i\Phi/2} & 0 \\ 0 & e^{-i\Phi/2} \end{pmatrix}. \tag{1.48}$$

Since Λ, Ψ, and Φ are phase factors, we can freely tune them in experiments. So we can set $\Lambda = \Psi = \Phi = 0$ without loss of generality:

$$\underline{B} = \begin{pmatrix} \cos(\Theta/2) & \sin(\Theta/2) \\ -\sin(\Theta/2) & \cos(\Theta/2) \end{pmatrix}. \tag{1.49}$$

In experiments we have the following relations for transmissivity T and reflectivity R of the beam splitter:

$$\sqrt{T} = \cos(\Theta/2), \tag{1.50}$$

$$\sqrt{R} = -\sin(\Theta/2), \tag{1.51}$$

$$T + R = 1. \tag{1.52}$$

Note that we can determine the value of Θ from experiments. So the beam splitter matrix can be described as follows:

$$\underline{B} = \begin{pmatrix} \sqrt{T} & -\sqrt{R} \\ \sqrt{R} & \sqrt{T} \end{pmatrix}. \tag{1.53}$$

Although we did not explicitly say that we were using the Heisenberg picture thus far, we will switch to the Schrödinger picture from here on. It is sometime convenient to use the state-transformation picture of beam splitters. For that purpose we introduce the operators $\hat{L}_i$ ($i = 0, 1, 2, 3$) as follows:

$$\hat{L}_0 = \frac{1}{2}\left(\hat{a}_1^\dagger \hat{a}_1 + \hat{a}_2^\dagger \hat{a}_2\right), \tag{1.54}$$

$$\hat{L}_1 = \frac{1}{2}\left(\hat{a}_1^\dagger \hat{a}_2 + \hat{a}_1 \hat{a}_2^\dagger\right), \tag{1.55}$$

$$\hat{L}_2 = \frac{1}{2i}\left(\hat{a}_1^\dagger \hat{a}_2 - \hat{a}_1 \hat{a}_2^\dagger\right), \tag{1.56}$$

$$\hat{L}_3 = \frac{1}{2}\left(\hat{a}_1^\dagger \hat{a}_1 - \hat{a}_2^\dagger \hat{a}_2\right). \tag{1.57}$$

These operators have the following relations:

$$e^{-i\Lambda\hat{L}_0}\begin{pmatrix}\hat{a}_1\\\hat{a}_2\end{pmatrix}e^{i\Lambda\hat{L}_0} = e^{i\frac{\Lambda}{2}}\begin{pmatrix}\hat{a}_1\\\hat{a}_2\end{pmatrix}, \tag{1.58}$$

$$e^{-i\Theta\hat{L}_2}\begin{pmatrix}\hat{a}_1\\\hat{a}_2\end{pmatrix}e^{i\Theta\hat{L}_2} = \begin{pmatrix}\cos(\Theta/2) & \sin(\Theta/2)\\ -\sin(\Theta/2) & \cos(\Theta/2)\end{pmatrix}\begin{pmatrix}\hat{a}_1\\\hat{a}_2\end{pmatrix}, \tag{1.59}$$

$$e^{-i\Phi\hat{L}_3}\begin{pmatrix}\hat{a}_1\\\hat{a}_2\end{pmatrix}e^{i\Phi\hat{L}_3} = \begin{pmatrix}e^{i\frac{\Phi}{2}} & 0\\ 0 & e^{-i\frac{\Phi}{2}}\end{pmatrix}\begin{pmatrix}\hat{a}_1\\\hat{a}_2\end{pmatrix}. \tag{1.60}$$

From a comparison between Eq. (1.48) and the time-evolution relationship (Eq. (1.8))

$$\hat{A}(t) = e^{i\frac{\hat{H}}{\hbar}t}\hat{A}(0)e^{-i\frac{\hat{H}}{\hbar}t}, \tag{1.61}$$

a beam splitter operator $\hat{B}$ can be described as follows:

$$\hat{B} = e^{i\Phi\hat{L}_3}e^{i\Theta\hat{L}_2}e^{i\Psi\hat{L}_3}e^{i\Lambda\hat{L}_0}. \tag{1.62}$$

Here we take $\hat{L}_i$ as a beam-splitter Hamiltonian and $-\Lambda$, $-\Psi$, $-\Phi$, and $-\Theta$ as time.

Now we got everything we need for the explanation of balanced homodyne measurement. Let's get back there.

1.3.2 Balanced Homodyne Measurement

We use a 50/50 beam splitter for balanced homodyne measurement, i.e., transmissivity and reflectivity, are both 0.5 ($T = R = 0.5$ in Eq. (1.53)). So the beam splitter matrix is

$$\underline{B} = \begin{pmatrix}1/\sqrt{2} & -1/\sqrt{2}\\ 1/\sqrt{2} & 1/\sqrt{2}\end{pmatrix}. \tag{1.63}$$

We will calculate the output of the balanced homodyne measurement $\hat{I}_2 - \hat{I}_1$ shown in Fig. 1.4 by using above matrix. Here the output $\hat{I}_2 - \hat{I}_1$ is the photocurrent difference of two detectors and $\hat{I}_1$ and $\hat{I}_2$ correspond to the photon numbers of modes 1 and 2, i.e., $\hat{I}_1 = \hat{a}_1'^\dagger \hat{a}_1'$ and $\hat{I}_2 = \hat{a}_2'^\dagger \hat{a}_2'$.

From Eq. (1.63)

$$\hat{I}_1 = \hat{a}_1'^\dagger \hat{a}_1'$$
$$= \frac{1}{2}\left(\hat{a}_1^\dagger \hat{a}_1 + \hat{a}_2^\dagger \hat{a}_2 - \hat{a}_1^\dagger \hat{a}_2 - \hat{a}_1 \hat{a}_2^\dagger\right), \tag{1.64}$$

and

$$\hat{I}_2 = \hat{a}_2'^\dagger \hat{a}_2'$$
$$= \frac{1}{2}\left(\hat{a}_1^\dagger \hat{a}_1 + \hat{a}_2^\dagger \hat{a}_2 + \hat{a}_1^\dagger \hat{a}_2 + \hat{a}_1 \hat{a}_2^\dagger\right). \tag{1.65}$$

So the output of the balanced homodyne measurement $\hat{I}_2 - \hat{I}_1$ can be calculated as

$$\hat{I}_2 - \hat{I}_1 = \hat{a}_1^\dagger \hat{a}_2 + \hat{a}_1 \hat{a}_2^\dagger. \tag{1.66}$$

When the input states of light beams 1 and 2 are $|\psi\rangle_1$ and $|\varphi\rangle_2$, which means that the overall state of the input is $|\psi\rangle_1 \otimes |\varphi\rangle_2$, the expected value of the output of the balanced homodyne measurement can be calculated as

$$\langle \hat{I}_2 - \hat{I}_1 \rangle = {}_2\langle\varphi| \otimes {}_1\langle\psi|(\hat{I}_2 - \hat{I}_1)|\psi\rangle_1 \otimes |\varphi\rangle_2$$
$$= {}_2\langle\varphi| \otimes {}_1\langle\psi|(\hat{a}_1^\dagger \hat{a}_2 + \hat{a}_1 \hat{a}_2^\dagger)|\psi\rangle_1 \otimes |\varphi\rangle_2. \tag{1.67}$$

Here one of the input light beams $\hat{a}_2$ is a local oscillator beam which is in a coherent state $|\alpha\rangle$ with a large amplitude $|\alpha|$. Then

$$\langle \hat{I}_2 - \hat{I}_1 \rangle = {}_2\langle\alpha| \otimes {}_1\langle\psi|(\hat{a}_1^\dagger \hat{a}_2 + \hat{a}_1 \hat{a}_2^\dagger)|\psi\rangle_1 \otimes |\alpha\rangle_2$$
$$= {}_1\langle\psi|\hat{a}_1^\dagger|\psi\rangle_1 \cdot {}_2\langle\alpha|\hat{a}_2|\alpha\rangle_2$$
$$\quad + {}_1\langle\psi|\hat{a}_1|\psi\rangle_1 \cdot {}_2\langle\alpha|\hat{a}_2^\dagger|\alpha\rangle_2$$
$$= {}_1\langle\psi|\hat{a}_1^\dagger|\psi\rangle_1\alpha + {}_1\langle\psi|\hat{a}_1|\psi\rangle_1\alpha^*. \tag{1.68}$$

When we use $\alpha = |\alpha|e^{i\theta}$ in the equation above, we get

$$\langle \hat{I}_2 - \hat{I}_1 \rangle = 2|\alpha|\left({}_1\langle\psi|\hat{x}_1|\psi\rangle_1 \cos\theta + {}_1\langle\psi|\hat{p}_1|\psi\rangle_1 \sin\theta\right), \tag{1.69}$$

where we use $\hat{a}_1 = \hat{x}_1 + i\hat{p}_1$.

From Eq. (1.69) we can see that the output of the balanced homodyne measurement is proportional to the inner product between the expected-value vector of $({}_1\langle\psi|\hat{x}_1|\psi\rangle_1, {}_1\langle\psi|\hat{p}_1|\psi\rangle_1)$ and the local-oscillator-amplitude vector of $(|\alpha|\cos\theta, |\alpha|\sin\theta)$. Moreover, since the amplitude $|\alpha|$ of the local oscillator is usually large, we can amplify a very weak input like a single photon.

Figure 1.6 shows an interpretation of balanced homodyne measurement output. A balanced homodyne measurement corresponds to projection of an input to a local oscillator beam.

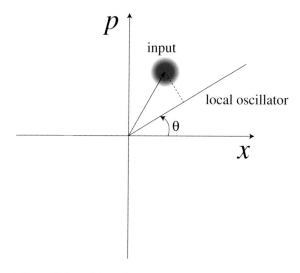

Fig. 1.6 Interpretation of balanced homodyne measurement output. The output of the balanced homodyne measurement is proportional to the inner product between the expected value vector of $(_1\langle\psi|\hat{x}_1|\psi\rangle_1, _1\langle\psi|\hat{p}_1|\psi\rangle_1)$ and that of local oscillator amplitude of $(|\alpha|\cos\theta, |\alpha|\sin\theta)$. So it corresponds to projection of an input to the axis at the local oscillator phase. Of course, the input is not limited by a coherent state

So far we have discussed the expected value of balanced homodyne measurements. We will from now on discuss the instantaneous value of balanced homodyne measurement of $\hat{I}_2 - \hat{I}_1$. Precisely speaking the local oscillator amplitude $|\alpha|$ is an averaged value. However, since it is usually much bigger than the quantum fluctuation $\sqrt{\langle\Delta\hat{x}_2^2\rangle} = \sqrt{\langle\Delta\hat{p}_2^2\rangle} = 1/2$, it can be treated as constant. So the instantaneous value of balanced homodyne measurement corresponds to the instantaneous value of $\hat{x}_1\cos\theta + \hat{p}_1\sin\theta$, namely, the instantaneous value of projection of the input onto the axis at the local oscillator phase (x_θ). By using this method we can get the distribution of the measured value for the component of the input at the local oscillator phase θ. When we scan θ, we can get the phase dependence which corresponds to the temporal structure of the input electric field. Figure 1.7 shows this example. It is phase dependence of amplitude for very weak laser light.

We will discuss these things more precisely in the next section especially on eigenstates of quadrature amplitude operators and marginal distributions.

1.3.3 Eigenstates of Quadrature Amplitude Operators and Marginal Distributions

As mentioned in Sect. 1.1, quadrature amplitude operators $\hat{x}$ and $\hat{p}$ ($\hat{a} = \hat{x} + i\hat{p}$, $\hbar = 1/2$) correspond to sine and cosine components of the electric field of light and they are canonical conjugate variables which satisfy $[\hat{x}, \hat{p}] = i/2$. So they correspond in

Fig. 1.7 Phase dependence of amplitude for very weak laser light

some sense to "position" and "momentum". Furthermore we can define eigenstates of quadrature amplitude operators $|x\rangle$ and $|p\rangle$ just like those of position and momentum operators as

$$\hat{x}|x\rangle = x|x\rangle, \tag{1.70}$$

$$\hat{p}|p\rangle = p|p\rangle. \tag{1.71}$$

Here they satisfy the following relations:

$$\langle x|x'\rangle = \delta(x - x'), \tag{1.72}$$

$$\langle p|p'\rangle = \delta(p - p'), \tag{1.73}$$

$$\int_{-\infty}^{+\infty} dx\,|x\rangle\langle x| = \hat{I}, \tag{1.74}$$

$$\int_{-\infty}^{+\infty} dp\,|p\rangle\langle p| = \hat{I}, \tag{1.75}$$

$$\langle x|p\rangle = \frac{1}{\sqrt{\pi}}e^{i2xp}. \tag{1.76}$$

Since the input is projected onto the x_θ axis in balanced homodyne measurements, we have to define the operator $\hat{x}_\theta$ as

$$\hat{x}_\theta = \hat{x}\cos\theta + \hat{p}\sin\theta. \tag{1.77}$$

The eigenstate $|x_\theta\rangle$ should satisfy

$$\hat{x}_\theta|x_\theta\rangle = x_\theta|x_\theta\rangle. \tag{1.78}$$

Here,

$$\langle x_\theta | x_\theta' \rangle = \delta(x_\theta - x_\theta'), \tag{1.79}$$

$$\int_{-\infty}^{+\infty} dx_\theta \, |x_\theta\rangle \langle x_\theta| = \hat{I}. \tag{1.80}$$

Projection onto the x_θ axis corresponds to taking inner product with $|x_\theta\rangle$ and the probability distribution can be calculated by squaring the inner product. More precisely, when the input state is $|\psi\rangle$, the inner product is $\langle x_\theta|\psi\rangle$ and it is equal to the wave function $\psi(x_\theta)$. Of course, the probability distribution is $|\psi(x_\theta)|^2$.

Let's think about balanced homodyne measurement of a coherent state $|\alpha\rangle$. The probability distribution should be $|\langle x_\theta|\alpha\rangle|^2$ from above discussion. It is called a marginal distribution. We will calculate it here, but before doing so, we have to make some preparations.

First of all we define $\hat{a}_\theta$ here. That is

$$\hat{a}_\theta = \hat{x}_\theta + i\hat{p}_\theta. \tag{1.81}$$

$\hat{x}_\theta$ is already defined as in Eq. (1.77) and $\hat{p}_\theta$ is defined as follows:

$$\hat{p}_\theta = -\hat{x}\sin\theta + \hat{p}\cos\theta. \tag{1.82}$$

From these definitions, we can get

$$\hat{a}_\theta = e^{-i\theta}\hat{a}. \tag{1.83}$$

This can be verified as follows:

$$\begin{aligned}
\hat{a}_\theta &= \hat{x}_\theta + i\hat{p}_\theta \\
&= (\hat{x}\cos\theta + \hat{p}\sin\theta) + i(-\hat{x}\sin\theta + \hat{p}\cos\theta) \\
&= \hat{x}(\cos\theta - i\sin\theta) + \hat{p}(\sin\theta + i\cos\theta) \\
&= \frac{\hat{a} + \hat{a}^\dagger}{2}(\cos\theta - i\sin\theta) + \frac{\hat{a} - \hat{a}^\dagger}{2i}(\sin\theta + i\cos\theta) \\
&= e^{-i\theta}\hat{a}.
\end{aligned} \tag{1.84}$$

The commutation relationship of $\hat{x}_\theta$ and $\hat{p}_\theta$ is $[\hat{x}_\theta, \hat{p}_\theta] = i/2$ ($\hbar = 1/2$) which is exactly the same as that of $\hat{x}$ and $\hat{p}$. It can be verified as follows:

$$\begin{aligned}
[\hat{x}_\theta, \hat{p}_\theta] &= \hat{x}_\theta\hat{p}_\theta - \hat{p}_\theta\hat{x}_\theta \\
&= (\hat{x}\cos\theta + \hat{p}\sin\theta)(-\hat{x}\sin\theta + \hat{p}\cos\theta) \\
&\quad - (-\hat{x}\sin\theta + \hat{p}\cos\theta)(\hat{x}\cos\theta + \hat{p}\sin\theta)
\end{aligned}$$

$$= \hat{x}\hat{p}\cos^2\theta - \hat{p}\hat{x}\sin^2\theta + \hat{x}\hat{p}\sin^2\theta - \hat{p}\hat{x}\cos^2\theta$$
$$= \hat{x}\hat{p} - \hat{p}\hat{x}$$
$$= \frac{i}{2}. \tag{1.85}$$

Now we are ready to calculate $|\langle x_\theta | \alpha \rangle|^2$. We first get an equation which $\langle x_\theta | \alpha \rangle$ should satisfy. From Eq. (1.84) and $\hat{a}|\alpha\rangle = \alpha|\alpha\rangle$

$$\langle x_\theta | \hat{a}_\theta | \alpha \rangle = \langle x_\theta | e^{-i\theta} \hat{a} | \alpha \rangle$$
$$= e^{-i\theta} \alpha \langle x_\theta | \alpha \rangle. \tag{1.86}$$

With $\hat{a}_\theta = \hat{x}_\theta + i\hat{p}_\theta$ it becomes

$$\langle x_\theta | \hat{a}_\theta | \alpha \rangle = \langle x_\theta | (\hat{x}_\theta + i\hat{p}_\theta) | \alpha \rangle$$
$$= \hat{x}_\theta \langle x_\theta | \alpha \rangle + i \langle x_\theta | \hat{p}_\theta | \alpha \rangle$$
$$= \hat{x}_\theta \langle x_\theta | \alpha \rangle + \frac{1}{2} \frac{d}{dx} \langle x_\theta | \alpha \rangle. \tag{1.87}$$

Here we used the relation

$$\langle x_\theta | \hat{p}_\theta | \psi \rangle = -i \frac{1}{2} \frac{d}{dx_\theta} \langle x_\theta | \psi \rangle. \tag{1.88}$$

By combining Eqs. (1.86) and (1.87) we see that the equation which $\langle x_\theta | \alpha \rangle$ should satisfy is

$$(x_\theta - e^{-i\theta}\alpha)\langle x_\theta | \alpha \rangle + \frac{1}{2} \frac{d}{dx_\theta} \langle x_\theta | \alpha \rangle = 0. \tag{1.89}$$

By solving this equation we get

$$\langle x_\theta | \alpha \rangle = c e^{-x_\theta^2 + 2\alpha e^{-i\theta} x_\theta}, \tag{1.90}$$

where c is a normalization constant. By using a normalization condition $\int_{-\infty}^{+\infty} dx_\theta$ $|\langle x_\theta | \alpha \rangle|^2 = 1$ and $\int dx\, e^{-x^2} = \sqrt{\pi}$ we get

$$c = \left(\frac{2}{\pi}\right)^{\frac{1}{4}} e^{-(\alpha_R \cos\theta + \alpha_I \sin\theta)^2}, \tag{1.91}$$

where $\alpha = \alpha_R + i\alpha_I$.

So the marginal distribution of coherent state $|\langle x_\theta | \alpha \rangle|^2$ can be calculated as

$$|\langle x_\theta | \alpha \rangle|^2 = \sqrt{\frac{2}{\pi}} e^{-2[x_\theta - (\alpha_R \cos\theta + \alpha_I \sin\theta)]^2}. \tag{1.92}$$

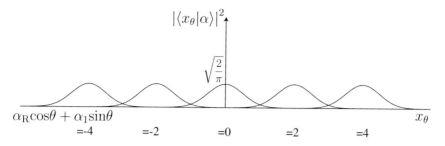

Fig. 1.8 Marginal distribution of a coherent state. The peak position changes as the local oscillator phase θ changes with the relation of $\alpha_R \cos \theta + \alpha_I \sin \theta$

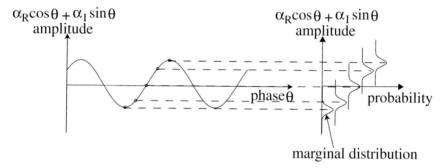

Fig. 1.9 Relation between the marginal distribution and phase dependence of amplitude shown in Fig. 1.7

It is a Gaussian distribution with a variance of 1/2 and the peak is at $\alpha_R \cos \theta + \alpha_I \sin \theta$, which is shown in Fig. 1.8. As shown in the figure the peak position changes with the expression $\alpha_R \cos \theta + \alpha_I \sin \theta$ which depends on the local oscillator phase θ. We also show the relation between the marginal distribution and Fig. 1.7 as Fig. 1.9.

As a final subject in this section we will discuss a vacuum state $|0\rangle$ as a special case of a coherent state $|\alpha\rangle$. We can get the marginal distribution of a vacuum when we set $\alpha = \alpha_R + i\alpha_I = 0$ in Eq. (1.92) as

$$|\langle x_\theta|0\rangle|^2 = \sqrt{\frac{2}{\pi}} e^{-2x_\theta^2}. \tag{1.93}$$

This is shown in Fig. 1.10. The important point here is that the peak position does not change as the local oscillator phase θ changes. So when we make a balanced homodyne measurement for a vacuum state $|0\rangle$, we get the phase dependence of amplitude as shown in Fig. 1.11.

In classical mechanics a vacuum means nothing, but we can see something in Fig. 1.11. In quantum mechanics we regard an annihilation operator $\hat{a} = \hat{x} + i\hat{p}$ as the quantum complex amplitude and $\hat{x}$ and $\hat{p}$ as sine and cosine components of the electric field of light. Moreover, $\hat{x}$ and $\hat{p}$ satisfy $[\hat{x}, \hat{p}] = i/2$ ($\hbar = 1/2$) and they have

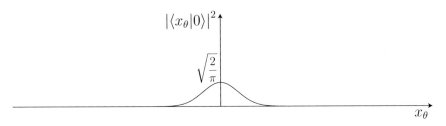

Fig. 1.10 Marginal distribution of a vacuum state $|0\rangle$. The peak position does not change as the local oscillator phase θ changes

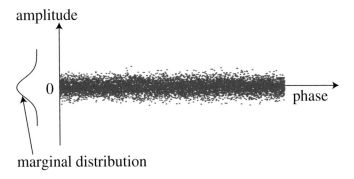

Fig. 1.11 Phase dependence of amplitude of a vacuum state $|0\rangle$

the uncertainty relationship as shown in Eq. (1.19). So sine and cosine components of the electric field of light cannot be determined simultaneously. In other words, when one of them is determined then the other becomes completely undetermined. In the case of a vacuum, it is a symmetric and minimum-uncertainty state, for which the equality of Eq. (1.19) holds. Now it is obvious that the phase dependence of amplitude of a vacuum state $|0\rangle$ becomes like Fig. 1.11. It comes from the uncertainty principle. The energy comes from $\hbar\omega/2$ of $E_n = \hbar\omega(n + 1/2)$ which corresponds to zero-point fluctuation of a quantum.

As shown in this section we can visualize a quantum state of light by using balanced homodyne measurements like Figs. 1.7 and 1.11. In the next sections we will show nonclassical states of light by using this method.

1.4 Single-Photon States

Some readers might think it is weird when they hear that a single-photon state is a nonclassical state. A lot of readers might think that they can get a single-photon state when they reduce laser power by using a neutral density filter. Unfortunately it is not true. Very weak laser light and a single-photon state $|1\rangle$ are totally different.

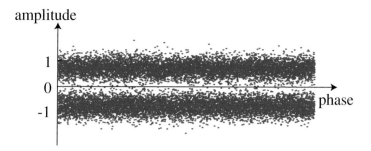

Fig. 1.12 Phase dependence of amplitude of a single-photon state $|1\rangle$

In the case of very weak laser light, when we count the photon number, it is mostly zero and sometimes one. That means the averaged photon number is much less than one. On the other hand, in the case of a sigle-photon state $|1\rangle$ the count is *always* one. It is a phenomenon totally different from the classical world.

The phase dependence of amplitude of very weak laser light is shown in Fig. 1.7. How does that of a single-photon state $|1\rangle$ look like? The answer is shown in Fig. 1.12. The big difference between Figs. 1.7 and 1.12 is that there is a wavy structure in Fig. 1.7 but there isn't in Fig. 1.12. What is going on here?

Since a single-photon state $|1\rangle$ is an energy eigenstate which means the energy is determined, the conjugate variable, i.e., the phase, should be totally undetermined in accordance with the uncertainty principle. So a single photon does not have a wave structure and it is interpreted as a wavelet without a phase, which is very nonclassical! Therefore we cannot make a single photon state $|1\rangle$ with reducing the power of laser light, because very weak laser light always has to have a wave structure.

So far we have given an intuitive explanation of the single-photon state $|1\rangle$. In the next section we will explain it with equations. Especially we will show the reason why the phase dependence of amplitude of a single-photon state $|1\rangle$ looks like Fig. 1.12.

1.4.1 Marginal Distribution of a Single-Photon State

We will calculate the marginal distribution of a single-photon state $|\langle x_\theta|1\rangle|^2$ here. For that purpose, we should rewrite the equation of $\langle x_\theta|0\rangle$ and $\langle x_\theta|1\rangle$ in the following way:

$$
\begin{aligned}
\langle x_\theta|1\rangle &= \langle x_\theta|\hat{a}^\dagger|0\rangle \\
&= \langle x_\theta|e^{i\theta}\hat{a}_\theta^\dagger|0\rangle \\
&= \langle x_\theta|e^{i\theta}(\hat{x}_\theta - i\hat{p}_\theta)|0\rangle \\
&= e^{i\theta}x_\theta\langle x_\theta|0\rangle - e^{i\theta}\frac{1}{2}\frac{d}{dx_\theta}\langle x_\theta|0\rangle,
\end{aligned} \tag{1.94}
$$

where we used Eqs. (1.84) and (1.88). Since a vacuum state $|0\rangle$ is a special case of a coherent state $|\alpha\rangle$ with $\alpha = 0$, we can use Eqs. (1.90) and (1.91). Then

$$\langle x_\theta|0\rangle = \left(\frac{2}{\pi}\right)^{\frac{1}{4}} e^{-x_\theta^2}. \qquad (1.95)$$

By using this result and Eq. (1.94) we get

$$\langle x_\theta|1\rangle = 2\left(\frac{2}{\pi}\right)^{\frac{1}{4}} x_\theta e^{-x_\theta^2}. \qquad (1.96)$$

So the marginal distribution $|\langle x_\theta|1\rangle|^2$ can be calculated with the following expression:

$$|\langle x_\theta|1\rangle|^2 = 4\sqrt{\frac{2}{\pi}} x_\theta^2 e^{-2x_\theta^2}. \qquad (1.97)$$

We plot it and get Fig. 1.13. The important point here is that the shape does not change even when the local oscillator phase changes, which is similar to the case of a vacuum state $|0\rangle$ but is totally different from the case of a coherent state. By using Fig. 1.13 we can understand Fig. 1.12 as shown in Fig. 1.14.

Let's think about the physical meaning of Figs. 1.12 and 1.14. A single photon should be an electromagnetic wave. We simply cannot see the wave structure. It is because a single-photon state $|1\rangle$ is an energy eigenstate which means that the energy is determined, and the conjugate variable, i.e., the phase, should be totally undetermined in accordance with the uncertainty principle as pointed out in the previous section. So the image might be like Fig. 1.15, which means a total mixture of waves with a constant amplitude but random phases.

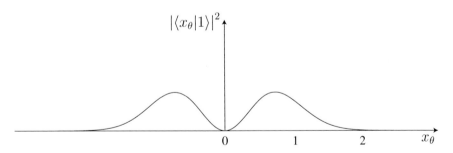

Fig. 1.13 Marginal distribution of a single-photon state $|1\rangle$. The shape does not change even when the local oscillator phase changes, which is similar to the case of a vacuum state $|0\rangle$ but is totally different from the case of a coherent state. Note that in this book we use $\hbar = 1/2$ and $\hat{a} = \hat{x} + i\hat{p}$ when we do not make a special remark. So the peak of the marginal distribution does not exist at $x_\theta = 1$ but $x_\theta = 1/\sqrt{2}$. On the other hand, when we use $\hbar = 1$ and $\hat{a} = (\hat{x} + i\hat{p})/\sqrt{2}$, the peak exists at $x_\theta = 1$. Therefore some people like $\hbar = 1$ and $\hat{a} = (\hat{x} + i\hat{p})/\sqrt{2}$. However the author likes $\hbar = 1/2$ and $\hat{a} = \hat{x} + i\hat{p}$. It is because the author hates the $\sqrt{2}$ in $\hat{a} = (\hat{x} + i\hat{p})/\sqrt{2}$. Of course there is no fundamental difference between them. You can use which one you like

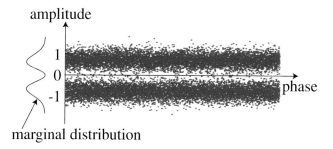

amplitude

phase

marginal distribution

Fig. 1.14 Marginal distribution and phase dependence of amplitude of a single-photon state $|1\rangle$. The shape of the marginal distribution does not change even when the local oscillator phase θ changes

Fig. 1.15 An image of a single photon state $|1\rangle$. It is a total mixture of waves with a constant amplitude but random phases

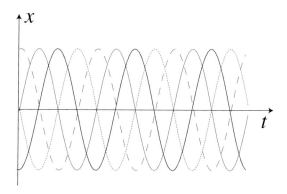

1.5   Photon-Number States

The story in the previous section is not only for light. General harmonic oscillators have the same property. So we can visualize general harmonic oscillators with the same method.

Everybody knows that the wave function of a harmonic oscillator $\varphi_n(x)$ can be described as

$$\varphi_n(x) = \sqrt{\frac{1}{2^n n!}} \sqrt{\frac{m\omega}{\pi\hbar}} H_n\left(\sqrt{\frac{m\omega}{\hbar}} x\right) e^{-\frac{m\omega}{2\hbar} x^2}, \tag{1.98}$$

where n is a quantum number, $H_n(x)$ is the Hermite polynomials, m is the mass, and ω is the angular frequency of the harmonic oscillation. However, almost nobody knows that $|\varphi_1(x)|^2$ corresponds to the marginal distribution of a single-photon state $|1\rangle$ and that the phase dependence of amplitude is like Fig. 1.14, where we set $m = 1$, $\omega = 1$, and $\hbar = 1/2$. Similarly $|\varphi_0(x)|^2$ corresponds to the marginal distribution of a vacuum.

This correspondence can be extended to general photon-number states of light $|n\rangle$ (eigenstates of $\hat{n} = \hat{a}^\dagger \hat{a}$, Fock states), namely, $|\varphi_n(x)|^2$ corresponds to the marginal distribution of photon-number states $|n\rangle$. Here we think about a two-photon state $|2\rangle$, a three-photon state $|3\rangle$, and a four-photon state $|4\rangle$.

In Eq. (1.98) we set $m = 1, \omega = 1, \hbar = 1/2$ and by using the Hermite polynomials,

$$H_0(\xi) = 1, \tag{1.99}$$

$$H_1(\xi) = 2\xi, \tag{1.100}$$

$$H_2(\xi) = 4\xi^2 - 2, \tag{1.101}$$

$$H_3(\xi) = 8\xi^3 - 12\xi, \tag{1.102}$$

$$H_4(\xi) = 16\xi^4 - 48\xi^2 + 12, \tag{1.103}$$

we can get

$$\varphi_0(x) = \left(\frac{2}{\pi}\right)^{\frac{1}{4}} e^{-x^2}, \tag{1.104}$$

$$\varphi_1(x) = 2\left(\frac{2}{\pi}\right)^{\frac{1}{4}} xe^{-x^2}, \tag{1.105}$$

$$\varphi_2(x) = \left(\frac{1}{2\pi}\right)^{\frac{1}{4}} (4x^2 - 1)e^{-x^2}, \tag{1.106}$$

$$\varphi_3(x) = \frac{1}{\sqrt{3}}\left(\frac{1}{2\pi}\right)^{\frac{1}{4}} (x^3 - 6x)e^{-x^2}, \tag{1.107}$$

$$\varphi_4(x) = \frac{1}{2\sqrt{3}}\left(\frac{1}{2\pi}\right)^{\frac{1}{4}} (16x^4 - 24x^2 + 3)e^{-x^2}. \tag{1.108}$$

The marginal distributions of these states will then be

$$|\varphi_0(x)|^2 = \sqrt{\frac{2}{\pi}} e^{-2x^2}, \tag{1.109}$$

$$|\varphi_1(x)|^2 = 4\sqrt{\frac{2}{\pi}} x^2 e^{-2x^2}, \tag{1.110}$$

$$|\varphi_2(x)|^2 = \sqrt{\frac{1}{2\pi}} (4x^2 - 1)^2 e^{-2x^2}, \tag{1.111}$$

$$|\varphi_3(x)|^2 = \frac{1}{3}\sqrt{\frac{1}{2\pi}} (x^3 - 6x)^2 e^{-2x^2}, \tag{1.112}$$

$$|\varphi_4(x)|^2 = \frac{1}{12}\sqrt{\frac{1}{2\pi}} (16x^4 - 24x^2 + 3)^2 e^{-2x^2}. \tag{1.113}$$

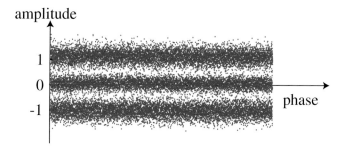

Fig. 1.16 Phase dependence of amplitude of a two-photon state $|2\rangle$

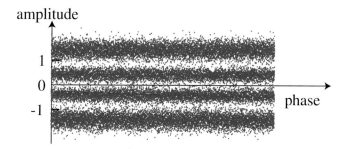

Fig. 1.17 Phase dependence of amplitude of a three-photon state $|3\rangle$

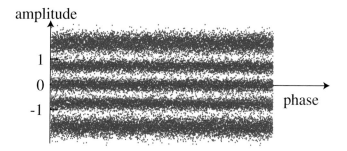

Fig. 1.18 Phase dependence of amplitude of a four-photon state $|4\rangle$

Similar to a vacuum state $|0\rangle$ and a single-photon state $|1\rangle$, these marginal distributions do not depend on the phase. These are phase-insensitive states, which mean that these are "waves" without wave structures. It is of course because these states are energy eigenstates with determined energies and it follows from the uncertainty principle that the phases become fully undetermined. The situation is exactly the same as the cases for a vacuum state $|0\rangle$ and a single-photon state $|1\rangle$.

We plot Eqs. (1.111)–(1.113) as marginal distributions of a two-photon state $|2\rangle$, a three-photon state $|3\rangle$, and a four-photon state $|4\rangle$ in Figs. 1.16, 1.17 and 1.18. There are no phase dependence in any of these states, which means they are "waves without phase". However, a superposition of such states like coherent states has phase and shows wave nature. We will think about the origin of the wave nature in the next section.

1.6 Superposition States of a Vacuum and a Single-Photon State

In the previous section we explained the nature of photon-number states $|n\rangle$. We showed that there is no phase dependence of amplitude of the states. In this section we explain how we can create waves from particles like single photons $|1\rangle$, more precisely, we explain how we can create waves with a superposition.

As explained in the previous section, a vacuum $|0\rangle$ and a single photon $|1\rangle$ do not have any phase dependence of amplitude, which is shown in Figs. 1.19 and 1.20. However, if we think of their superposition $(|0\rangle + |1\rangle)/\sqrt{2}$, the situation changes totally. Before the detailed explanation we think about an incoherent mixture of a vacuum $|0\rangle$ and a single photon $|1\rangle$, which is shown in Fig. 1.21. This is just a mixture of Figs. 1.19 and 1.20, and totally different from a superposition of a vacuum $|0\rangle$ and a single photon $|1\rangle$.

Figure 1.22 shows the amplitude dependence on phase for a superposition of a vacuum $|0\rangle$ and a single photon $|1\rangle$, $(|0\rangle + |1\rangle)/\sqrt{2}$. We can see phase dependence of the amplitude, i.e., a wave structure, which is different from an incoherent mixture of them shown in Fig. 1.21. Where does the wave structure come from? To think about it, we show another phase dependence of amplitude of a superposition of a vacuum $|0\rangle$ and a single photon $|1\rangle$ in Fig. 1.23, which is now $(|0\rangle - |1\rangle)/\sqrt{2}$. From the comparison between Figs. 1.22 and 1.23, we can see that the wave structures have opposite phases. So we can draw the following two conclusions:

Fig. 1.19 Phase dependence of amplitude of a vacuum $|0\rangle$ (Fig. 1.11)

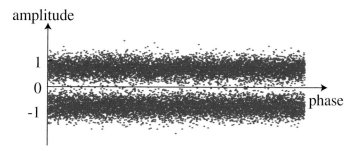

Fig. 1.20 Phase dependence of amplitude of a single-photon state $|1\rangle$ (Fig. 1.12)

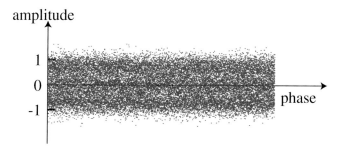

Fig. 1.21 Phase dependence of amplitude of an incoherent mixture of a vacuum $|0\rangle$ and a single photon $|1\rangle$

Fig. 1.22 Phase dependence of amplitude of a superposition of a vacuum $|0\rangle$ and a single photon $|1\rangle$, $(|0\rangle + |1\rangle)/\sqrt{2}$

1. The wave structure is created by a superposition of a vacuum $|0\rangle$ and a single photon $|1\rangle$. We can not create it with an incoherent mixture of them.
2. The phase of the wave structure is flipped when the sign of the superposition between a vacuum $|0\rangle$ and a single photon $|1\rangle$ changes.

From these facts we can say that the phase of the wave structure is determined by the relative phase between the vacuum state $|0\rangle$ and the single photon state $|1\rangle$. Moreover the average amplitude of the wave structure is $1/\sqrt{2}$ compared to the case of a single

Fig. 1.23 Phase dependence of amplitude of a superposition of a vacuum $|0\rangle$ and a single photon $|1\rangle$, $(|0\rangle - |1\rangle)/\sqrt{2}$

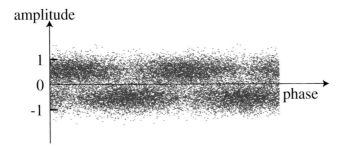

Fig. 1.24 Phase dependence of amplitude of a superposition of a vacuum $|0\rangle$ and a single photon $|1\rangle$, $(|0\rangle + 2|1\rangle)/\sqrt{5}$

photon $|1\rangle$. This reflects the factor of $1/\sqrt{2}$ of the superposition state $(|0\rangle \pm |1\rangle)/\sqrt{2}$. It is trivial in some sense because the probabilities of a vacuum and a single photon are $1/2$ respectively and the average photon number of the state should be a half, which corresponds to $1/\sqrt{2}$ of the average amplitude of a single photon.

To get more information on the influence of superposition factors on the amplitude of the wave structure, we think about $(|0\rangle + 2|1\rangle)/\sqrt{5}$. Figure 1.24 shows the phase dependence of amplitude of $(|0\rangle + 2|1\rangle)/\sqrt{5}$. The average amplitude of the wave structure is $2/\sqrt{5}$ compared to one for a single photon $|1\rangle$. This is also explained by the fact that the probabilities of a vacuum and a single photon are $1/5$ and $4/5$ respectively and the average photon number of the state should be $4/5$, which corresponds to $2/\sqrt{5}$ of the average amplitude of a single photon.

As another example we think about $(2|0\rangle + |1\rangle)/\sqrt{5}$. Figure 1.25 shows the phase dependence of amplitude of $(2|0\rangle + |1\rangle)/\sqrt{5}$. The average photon number is $0 \times 4/5 + 1 \times 1/5 = 1/5$ and the average amplitude of the wave structure should be $1/\sqrt{5}$ compared to the case of a single photon. We can check this in Fig. 1.25.

From above observations we found that the wave structure is created by a superposition of a vacuum $|0\rangle$ and a single photon $|1\rangle$. We also found that the phase and average amplitude are determined by the superposition factor. We have so far mysteriously explained the creation of waves from a superposition of a vacuum $|0\rangle$ and

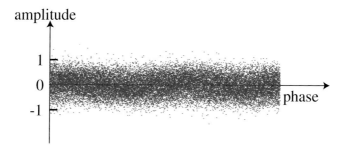

Fig. 1.25 Phase dependence of amplitude of a superposition of a vacuum $|0\rangle$ and a single photon $|1\rangle$, $(2|0\rangle + |1\rangle)/\sqrt{5}$

a single photon $|1\rangle$. However, it is trivial in some sense. You can easily understand it when you remember that a coherent state $|\alpha\rangle$ is a superposition of photon-number states:

$$|\alpha\rangle = e^{-\frac{|\alpha|^2}{2}} \sum_{n=0}^{\infty} \frac{\alpha^n}{\sqrt{n!}} |n\rangle. \tag{1.114}$$

When $|\alpha|$ is small enough, a coherent state can be regarded as a superposition of a vacuum $|0\rangle$ and a single photon $|1\rangle$. Of course, the wave structures of all types of superposition states of a vacuum $|0\rangle$ and a single photon $|1\rangle$ do not get a perfect sine curve. It is simply because we need the higher photon-number terms to create a perfect sine curve.

In the next section we will think about superposition of coherent states, which include all photon-number states.

1.7 Coherent States and Schrödinger Cat States

Before explaining superposition of coherent states, we add some explanation on coherent states. We can expand a coherent state as a superposition of photon-number states with Eq. (1.114), e.g., $\alpha = 1$:

$$
\begin{aligned}
|\alpha = 1\rangle &= e^{-\frac{1}{2}} \sum_{n=0}^{\infty} \frac{1}{\sqrt{n!}} |n\rangle \\
&= e^{-\frac{1}{2}} \left(|0\rangle + |1\rangle + \frac{1}{\sqrt{2}} |2\rangle + \frac{1}{\sqrt{6}} |3\rangle + \frac{1}{2\sqrt{6}} |4\rangle + \cdots \right). \tag{1.115}
\end{aligned}
$$

Figure 1.26 shows the phase dependence of amplitude of the coherent state $|\alpha = 1\rangle$. It is trivial that a coherent state has a wave structure. However, it seems strange that a superposition of phase-independent photon-number states has phase dependence.

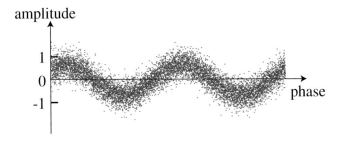

Fig. 1.26 Phase dependence of amplitude of a coherent state $|\alpha = 1\rangle$

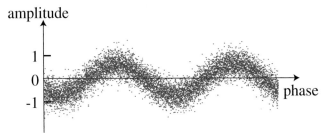

Fig. 1.27 Phase dependence of amplitude of a coherent state $|\alpha = -1\rangle$

Let us think about the opposite-phase coherent state $|\alpha = -1\rangle$. It can be expanded as a superposition of photon-number states with Eq. (1.114) as follows:

$$
|\alpha = -1\rangle = e^{-\frac{1}{2}} \sum_{n=0}^{\infty} \frac{(-1)^n}{\sqrt{n!}} |n\rangle
$$

$$
= e^{-\frac{1}{2}} \left(|0\rangle - |1\rangle + \frac{1}{\sqrt{2}}|2\rangle - \frac{1}{\sqrt{6}}|3\rangle + \frac{1}{2\sqrt{6}}|4\rangle - \cdots \right). \quad (1.116)
$$

Figure 1.27 shows the phase dependence of amplitude. It is trivial but the phase is flipped compared to Fig. 1.26.

From the comparison between Eqs. (1.115) and (1.116), we can see that the coefficients of even-photon-number states are the same but those of odd-photon-number states are sign-flipped, which correspond to that the phase of even-photon-number states are the same but those of odd-photon-number states are flipped. This situation is the same as that of a superposition of a vacuum $|0\rangle$ and a single photon $|1\rangle$ as shown in the previous section.

Now we move on to the explanation of a superposition of coherent states $N_{\alpha\pm}(|\alpha\rangle \pm |-\alpha\rangle)$ ($N_{\alpha\pm}$ is a normalization factor). This superposition state is called a Schrödinger's cat state. The reasons are the following. Coherent states are a superposition of many photons (quanta) as seen in Eq. (1.114). Our daily world or macroscopic world also consists of many atoms (quanta). In that sense coherent states can

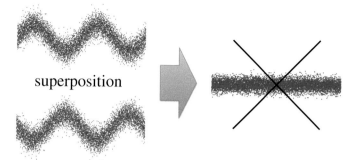

Fig. 1.28 Does a superposition of coherent states $|\alpha\rangle$ and $|-\alpha\rangle$ becomes a vacuum state?

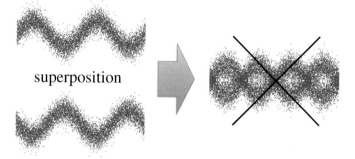

Fig. 1.29 Does a superposition of coherent states $|\alpha\rangle$ and $|-\alpha\rangle$ becomes a mixture of them?

be regarded as a macroscopic existence. Since a Schrödinger's cat state is a superposition state of macroscopic existence, i.e., a superposition of an alive and dead cat, a superposition of coherent states $|\alpha\rangle$ and $|-\alpha\rangle$ can be regarded as a Schrödinger's cat state. Here coherent states $|\alpha\rangle$ and $|-\alpha\rangle$ are phase-flipped semiclassical waves. Therefore there are some misunderstandings. Figure 1.28 shows one of them. Some people think that a superposition of coherent states $|\alpha\rangle$ and $|-\alpha\rangle$ become a vacuum state. Of course, it is not right. Figure 1.29 shows another misunderstanding. Some people think a superposition of coherent states $|\alpha\rangle$ and $|-\alpha\rangle$ becomes a mixture of them. Of course, that is not right, either. What is the true picture of the Schrödinger's cat state?

First we consider one of the Schrödinger cat's states, $N_{\alpha-}(|\alpha\rangle - |-\alpha\rangle)$. It is sometimes called a "minus cat state". We calculate $|\alpha\rangle - |-\alpha\rangle$ with Eq. (1.114) and get

$$|\alpha\rangle - |-\alpha\rangle = e^{-\frac{|\alpha|^2}{2}} \sum_{n=0}^{\infty} \frac{\alpha^n}{\sqrt{n!}} |n\rangle - e^{-\frac{|\alpha|^2}{2}} \sum_{n=0}^{\infty} \frac{(-\alpha)^n}{\sqrt{n!}} |n\rangle$$

$$= 2e^{-\frac{|\alpha|^2}{2}} \sum_{n=0}^{\infty} \frac{\alpha^{2n+1}}{\sqrt{(2n+1)!}} |2n+1\rangle. \qquad (1.117)$$

From this result we can see that this minus cat state is a superposition of odd photons. The normalizing factor will be calculated as Eq. (1.134), that is,

$$N_{\alpha-} = \frac{1}{\sqrt{2(1 - e^{-2|\alpha|^2})}}. \qquad (1.118)$$

So the minus cat state with $\alpha = 1$ is described as

$$N_{\alpha-}(|\alpha = 1\rangle - |\alpha = -1\rangle) = \sqrt{\frac{2e}{e^2 - 1}} \left(|1\rangle + \frac{1}{\sqrt{3!}} |3\rangle + \frac{1}{\sqrt{5!}} |5\rangle \cdots \right). \qquad (1.119)$$

Figure 1.30 shows the phase dependence of amplitude of the Schrödinger's cat state $N_{\alpha-}(|\alpha\rangle - |-\alpha\rangle)$ ($\alpha = 1$).

From a comparison between Figs. 1.29 and 1.30, we can see that a minus cat state or a superposition state of two coherent states is totally different from their mixture. Especially the zero probability around zero amplitude is the most prominent feature. As seen in Sect. 1.5, odd photon-number states show the zero probability around zero amplitude. Therefore the minus cat state also has this feature. In other words, quantum interference or superposition cancels the probability of zero amplitude. It corresponds to cancellation of even-photon-number components in Eq. (1.119).

Now we change the value of α from 1 to 2. Figure 1.31 shows the phase dependence of amplitude of the Schrödinger's cat state $N_{\alpha-}(|\alpha\rangle - |-\alpha\rangle)$ ($\alpha = 2$). It looks really complicated and it is the limit of this type of pictures. So we will introduce the Wigner function in the next section to handle this type of complicated situations.

Fig. 1.30 Phase dependence of amplitude of a Schrödinger's cat state $N_{\alpha-}(|\alpha\rangle - |-\alpha\rangle)$ ($\alpha = 1$)

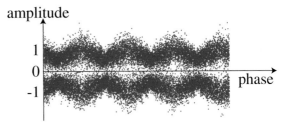

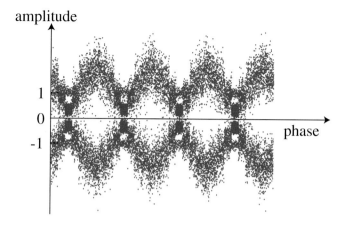

Fig. 1.31 Phase dependence of amplitude of a Schrödinger's cat state $N_{\alpha-}(|\alpha\rangle - |-\alpha\rangle)$ $(\alpha = 2)$

1.8   The Wigner Function

Until this section we have visualized quantum states of light with phase dependences of quadrature amplitudes. This is because we believe that wave structures appearing in coherent states give us an image of quantum states. However, as pointed out for a Schrödinger's cat state with $\alpha = 2$, there are some limitations to that method. So we introduce the Wigner function in this section to solve that problem.

The definition of the Wigner function $W(x, p)$ is

$$W(x, p) \equiv \frac{1}{2\pi\hbar} \int_{-\infty}^{\infty} d\xi \exp\left(-\frac{i}{\hbar}p\xi\right) \left\langle x + \frac{1}{2}\xi \middle| \hat{\rho} \middle| x - \frac{1}{2}\xi \right\rangle, \qquad (1.120)$$

where $\hat{\rho}$ is the density operator for the quantum state which we want to visualize. Here the definition of a density operator is

$$\hat{\rho} = \sum_{n=0}^{\infty} \rho_n |\psi_n\rangle\langle\psi_n|, \qquad (1.121)$$

and ρ_n is the probability of quantum state $|\psi_n\rangle$. For example, the density operator for a coherent state $|\alpha\rangle$ is $|\alpha\rangle\langle\alpha|$. Moreover, we can handle a mixed state as well as a pure state with the density operator, which cannot be written with a single $|\psi\rangle$. This feature is really powerful and we can handle any states with the Wigner function.

The Wigner function is called a pseudo-probability density function and it can be regarded as a probability distribution in phase space with "position" x and "momentum" p. Of course, x and p are quadrature amplitudes in quantum optics. The interpretation of the Wigner function is the following. The part $\langle x + \frac{1}{2}\xi| \hat{\rho} |x - \frac{1}{2}\xi\rangle$ in Eq. (1.120) represents the probability of a "quantum jump" from positions $(x - \frac{1}{2}\xi)$

to $\left(x + \frac{1}{2}\xi\right)$, and it can be regarded as the existence of the quantum at position x (a middle point of $x - \frac{1}{2}\xi$ and $x + \frac{1}{2}\xi$) and its movement by length ξ. The integral $\int_{-\infty}^{\infty} d\xi \exp\left(-\frac{i}{\hbar}p\xi\right)$ in Eq. (1.120) corresponds to the Fourier transformation from position space to momentum space and thus the length ξ is transformed to a momentum p. The Wigner function can therefore be regarded as a probability density distribution of x and p.

The above explanation gives us an impression that the Wigner function is just a classical probability distribution. However, it is not. It is because the Wigner function can be negative. As you know, a classical probability distribution cannot be negative. That is why the Wigner function is called a pseudo-probability density function.

The Wigner function has the following property:

$$\iint dx dp \, W(x, p) = 1. \tag{1.122}$$

This property corresponds to that a total probability becomes one. Moreover, two Wigner functions $|\psi_1\rangle$ and $|\psi_2\rangle$ have the following property when these Wigner functions correspond to two pure states $|\psi_1\rangle$ and $|\psi_2\rangle$, respectively:

$$|\langle \psi_1|\psi_2\rangle|^2 = 2\pi\hbar \iint dx dp \, W_1(x, p) W_2(x, p). \tag{1.123}$$

This property means that overlap between two Wigner functions corresponds to the overlap between two states. Naively speaking, we can say that two states are similar when the overlap between the two Wigner functions is big enough. Note that Eq. (1.123) is valid when the two states are pure states. In the case of mixed states, the situation is much more complicated.

We will show some examples of the Wigner functions. As the first example we will show the Wigner function of a vacuum $|0\rangle$. We put $|0\rangle\langle 0|$ to $\hat{\rho}$ in Eq. (1.120) and we use $\hbar = \frac{1}{2}$. Then we get the Wigner function of a vacuum $W_0(x, p)$ as follows:

$$W_0(x, p) = \frac{1}{\pi} \int d\xi e^{-2ip\xi} \langle x + \xi/2|0\rangle \langle 0|x - \xi/2\rangle. \tag{1.124}$$

To calculate this equation we first calculate $\langle x|0\rangle$. This is actually $\varphi_0(x)$ (Eq. (1.104)) calculated in Sect. 1.5, that is,

$$\psi_0(x) \equiv \langle x|0\rangle = \left(\frac{2}{\pi}\right)^{\frac{1}{4}} e^{-x^2}. \tag{1.125}$$

By using this we calculate Eq. (1.124).

$$W_0(x, p) = \frac{1}{\pi} \int d\xi e^{-2ip\xi} \langle x + \xi/2|0\rangle\langle 0|x - \xi/2\rangle$$

$$= \frac{1}{\pi} \int d\xi e^{-2ip\xi} \psi_0(x + \xi/2)\psi_0(x - \xi/2)$$

$$= \frac{1}{\pi} \int d\xi e^{-2ip\xi} \left(\frac{2}{\pi}\right)^{\frac{1}{4}} e^{-(x+\xi/2)^2} \left(\frac{2}{\pi}\right)^{\frac{1}{4}} e^{-(x-\xi/2)^2}$$

$$= \frac{2}{\pi} e^{-2(x^2+p^2)}. \tag{1.126}$$

From this result we see that the Wigner function of a vacuum $W_0(x, p)$ is a complex Gaussian distribution as shown in Fig. 1.32.

As the second example of the Wigner function we will take the one of a coherent state $|\alpha\rangle$ ($\alpha = x_0 + ip_0$) $W_\alpha(x, p)$. Since the density operator of a coherent state is $|\alpha\rangle\langle\alpha|$, we first calculate $\psi_\alpha(x) \equiv \langle x|\alpha\rangle$ to get $W_\alpha(x, p)$:

$$\psi_\alpha(x) = \langle x|\alpha\rangle,$$

$$= \langle x|\hat{D}(\alpha)|0\rangle$$

$$= \langle x|e^{-ix_0p_0}e^{i2p_0\hat{x}}e^{-i2x_0\hat{p}}|0\rangle$$

$$= e^{-ix_0p_0}e^{i2p_0x}\langle x|e^{-i2x_0\hat{p}}|0\rangle$$

$$= e^{-ix_0p_0}e^{i2p_0x}\psi_0(x - x_0)$$

$$= \left(\frac{2}{\pi}\right)^{\frac{1}{4}} e^{-ix_0p_0+i2p_0x-(x-x_0)^2}, \tag{1.127}$$

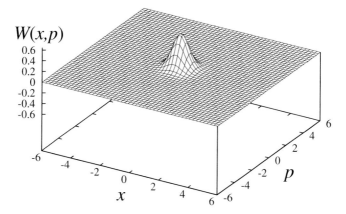

Fig. 1.32 The Wigner function of a vacuum $|0\rangle$

where $\hat{D}(\alpha)$ represents the displacement operation which will be used in the next chapter. We also used the following property:

$$
\begin{aligned}
\langle x|e^{-i2x_0\hat{p}}|0\rangle &= \langle x|\int_{-\infty}^{\infty} dp|p\rangle\langle p|e^{-i2x_0\hat{p}}|0\rangle \\
&= \int_{-\infty}^{\infty} dp\frac{1}{\pi}e^{i2xp-i2x_0p}\langle p|0\rangle \\
&= \int_{-\infty}^{\infty} dp\frac{1}{\pi}e^{i2p(x-x_0)}\langle p|0\rangle \\
&= \langle x-x_0|0\rangle.
\end{aligned}
\tag{1.128}
$$

As will be explained in the next chapter, the displacement operator $\hat{D}(\alpha)$ transforms a vacuum $|0\rangle$ to a coherent state $|\alpha\rangle$ and it can be regarded as laser oscillation. When $\alpha = x_0 + ip_0$, the displacement operator is

$$
\hat{D}(\alpha) = e^{-ix_0p_0}e^{i2p_0\hat{x}}e^{-i2x_0\hat{p}}.
\tag{1.129}
$$

By using Eq. (1.127) we get

$$
W_\alpha(x,p) = \frac{2}{\pi}e^{-2(x-x_0)^2-2(p-p_0)^2}.
\tag{1.130}
$$

We can see that the Wigner function of a coherent state $|\alpha\rangle$ ($\alpha = x_0 + ip_0$) is a shifted vacuum-Wigner function in phase space by (x_0, p_0) as shown in Fig. 1.33.

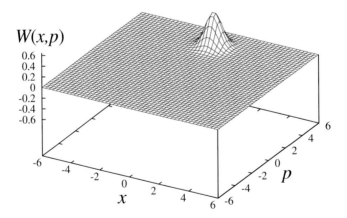

Fig. 1.33 The Wigner function of a coherent state $(x_0, p_0) = (0, 4)$

So far we have not seen the negativity of the Wigner function. However, we can see it in the one of a minus cat state. We will calculate the Wigner function of a minus cat state. First we calculate the normalization factor $N_{\alpha-}$ of the state vector of a minus cat state $|\text{cat}_{\alpha-}\rangle$. The minus cat state is defined as

$$|\text{cat}_{\alpha-}\rangle = N_{\alpha-}\left(|\alpha\rangle - |-\alpha\rangle\right). \tag{1.131}$$

The normalization condition is $\langle\text{cat}_{\alpha-}|\text{cat}_{\alpha-}\rangle = 1$ and we get

$$
\begin{aligned}
\langle\text{cat}_{\alpha-}|\text{cat}_{\alpha-}\rangle &= N_{\alpha-}^*\left(\langle\alpha| - \langle-\alpha|\right)N_{\alpha-}\left(|\alpha\rangle - |-\alpha\rangle\right) \\
&= |N_{\alpha-}|^2\left(\langle\alpha|\alpha\rangle + \langle-\alpha|-\alpha\rangle - \langle\alpha|-\alpha\rangle - \langle-\alpha|\alpha\rangle\right) \\
&= |N_{\alpha-}|^2\left(2 - \langle\alpha|-\alpha\rangle - \langle-\alpha|\alpha\rangle\right) \\
&= 1. \tag{1.132}
\end{aligned}
$$

By using the property of a coherent state $|\alpha\rangle$ given in Eq. (1.114), we get

$$
\begin{aligned}
\langle\alpha|-\alpha\rangle &= e^{\frac{-|\alpha|^2}{2}}\sum_{m=0}^{\infty}\frac{\alpha^{*m}}{\sqrt{m!}}\langle m| \cdot e^{\frac{-|\alpha|^2}{2}}\sum_{n=0}^{\infty}\frac{\alpha^n}{\sqrt{n!}}|n\rangle \\
&= e^{-|\alpha|^2}\sum_{n=0}^{\infty}\frac{(-|\alpha|^2)^n}{n!} \\
&= e^{-2|\alpha|^2}. \tag{1.133}
\end{aligned}
$$

Then by using Eq. (1.132) we get

$$N_{\alpha-} = \frac{1}{\sqrt{2(1 - e^{-2|\alpha|^2})}}. \tag{1.134}$$

Next, we get the density operator of a minus cat state $\hat{\rho}_{\text{cat}-}$. From the definition it should be

$$
\begin{aligned}
\hat{\rho}_{\text{cat}-} &= |N_{\alpha-}|^2\left(|\alpha\rangle - |-\alpha\rangle\right)\left(\langle\alpha| - \langle-\alpha|\right) \\
&= |N_{\alpha-}|^2\left(|\alpha\rangle\langle\alpha| + |-\alpha\rangle\langle-\alpha| - |\alpha\rangle\langle-\alpha| - |-\alpha\rangle\langle\alpha|\right). \tag{1.135}
\end{aligned}
$$

The Wigner function of a minus cat state $W_{\text{cat}\alpha-}(x, p)$ can therefore be derived from the definition of Eq. (1.120) as follows:

$$
\begin{aligned}
W_{\text{cat}\alpha-}(x, p) = \frac{1}{\pi}\int_{-\infty}^{\infty}d\xi e^{-2ip\xi}|N_{\alpha-}|^2\Big[&\langle x + \xi/2|\alpha\rangle\langle\alpha|x - \xi/2\rangle \\
&+ \langle x + \xi/2|-\alpha\rangle\langle-\alpha|x - \xi/2\rangle \\
&- \langle x + \xi/2|\alpha\rangle\langle-\alpha|x - \xi/2\rangle \\
&- \langle x + \xi/2|-\alpha\rangle\langle\alpha|x - \xi/2\rangle\Big]
\end{aligned}
$$

$$= \frac{1}{\pi} \int_{-\infty}^{\infty} d\xi e^{-2ip\xi} |N_{\alpha-}|^2 \Big[\langle x + \xi/2 | \alpha \rangle \langle \alpha | x - \xi/2 \rangle$$

$$+ \langle x + \xi/2 | -\alpha \rangle \langle -\alpha | x - \xi/2 \rangle$$

$$- \left(\frac{2}{\pi}\right)^{\frac{1}{4}} e^{-ix_0p_0+i2p_0(x+\xi/2)-(x+\xi/2-x_0)^2}$$

$$\times \left(\frac{2}{\pi}\right)^{\frac{1}{4}} e^{ix_0p_0+i2p_0(x-\xi/2)-(x-\xi/2-x_0)^2}$$

$$- \left(\frac{2}{\pi}\right)^{\frac{1}{4}} e^{-ix_0p_0-i2p_0(x+\xi/2)-(x+\xi/2-x_0)^2}$$

$$\times \left(\frac{2}{\pi}\right)^{\frac{1}{4}} e^{ix_0p_0-i2p_0(x-\xi/2)-(x-\xi/2-x_0)^2} \Big]$$

$$= |N_{\alpha-}|^2 W_\alpha(x,p) + |N_{\alpha-}|^2 W_\alpha(x,p)$$

$$- |N_{\alpha-}|^2 \frac{2}{\pi} \left[e^{i4(p_0x-x_0p)} + e^{-i4(p_0x-x_0p)} \right]$$

$$= \frac{1}{\pi(1 - e^{-2|\alpha|^2})} \Big[e^{-2(x-x_0)^2-2(p-p_0)^2} + e^{-2(x+x_0)^2-2(p+p_0)^2}$$

$$- 2e^{-2x^2-2p^2} \cos 4(p_0x - x_0p) \Big]. \tag{1.136}$$

From the equation above we can see that the Wigner function of a minus cat state $W_{\text{cat}\alpha-}$ has the structures of two coherent states of $|\alpha\rangle$ and $|-\alpha\rangle$, which are $e^{-2(x-x_0)^2-2(p-p_0)^2}$ and $e^{-2(x+x_0)^2-2(p+p_0)^2}$. On top of that we can see an interference term of $-2e^{-2x^2-2p^2} \cos 4(p_0x - x_0p)$. So we can say that an "alive cat state" $|\alpha\rangle$ and a "dead cat state" $|-\alpha\rangle$ exist simultaneously and the quantum interference is there. Moreover, the Wigner function has a minimum value at the origin $(x, p) = (0, 0)$ and the value is

$$W_{\text{cat}\alpha-}(0, 0) = \frac{e^{-2x_0^2-2p_0^2} + e^{-2x_0^2-2p_0^2} - 2}{\pi(1 - e^{-2|\alpha|^2})},$$

$$= -\frac{2}{\pi}. \tag{1.137}$$

It is always negative irrespective of the value of α.

Although the Wigner function is a sort of probability density function, it can take a negative value. Therefore it is called a "pseudo-"probability density function. The negative value of the Wigner function has no classical counterpart and it is a sign of nonclassicality. One can say that it is a nonclassical state when the state shows the negativity of the Wigner function. Figures 1.34 and 1.35 show the Wigner functions of minus cat states with $\alpha = 1$ and $\alpha = 2$, respectively.

The negative value of the Wigner function of a minus cat state with $\alpha = 1$ (Fig. 1.34) corresponds to the zero-probability around zero-amplitude in the phase dependence of amplitude of the state (Fig. 1.30). Of course, we already pointed out that there are two structures of coherent states $|\alpha\rangle$ and $|-\alpha\rangle$ in the Wigner function, which also appears in Fig. 1.30. Moreover, we saw a complicated structure in the phase dependence of amplitude of the minus cat state with $\alpha = 2$ in Fig. 1.31. We can see that it comes from the quantum interference term $-2e^{-2x^2-2p^2}\cos 4(p_0 x - x_0 p)$ in Eq. (1.136). The frequency of cosine changes depending on $\alpha = x_0 + ip_0$. In the case of $\alpha = 1$ it is one and in the case of $\alpha = 2$ it is two, which is two times "faster". Therefore the structure looks more complicated for $\alpha = 2$ than for $\alpha = 1$.

More quantitatively speaking the Wigner function $W(x, p)$ and the marginal distribution $|\psi(x_\theta)|^2$ ($x_\theta = x\cos\theta + p\sin\theta$) have the relation

$$|\psi(x_\theta)|^2 = \int_\infty^\infty dp_\theta W(x, p), \qquad (1.138)$$

where $dp_\theta = -x\sin\theta + p\cos\theta$. We can get the marginal distribution at the phase θ when we integrate the Wigner function along the axis which makes the angle θ with x axis. So we can get Figs. 1.30 and 1.31 from Figs. 1.34 and 1.35, respectively. Moreover, we can get a Wigner function from marginal distributions at various phase axes with the above relation, where the marginal distributions can be obtained by quantum (homodyne) tomography.

To check quantum coherence in a minus cat state, we will consider a mixed state of two coherent states $|\alpha\rangle$ and $|-\alpha\rangle$. The ability to handle such a mixed state is a

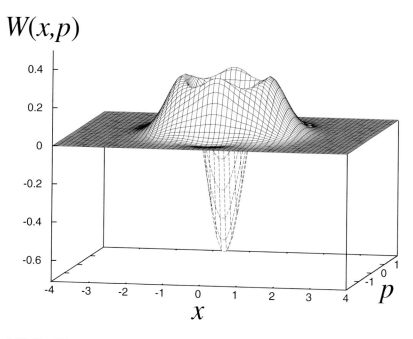

Fig. 1.34 The Wigner function of a minus cat state with $\alpha = 1$, where $(x_0, p_0) = (1, 0)$. We can see an "alive cat state" $|\alpha = 1\rangle$, a "dead cat state" $|\alpha = -1\rangle$ and the quantum interference with a negative value in between

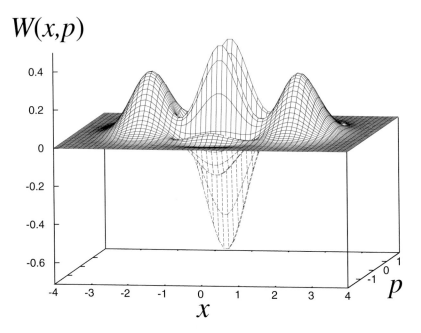

Fig. 1.35 The Wigner function of a minus cat state with $\alpha = 2$, where $(x_0, p_0) = (2, 0)$. We can see an "alive cat state" $|\alpha = 2\rangle$, a "dead cat state" $|\alpha = -2\rangle$ and the quantum interference with a negative value in between

big advantage of the Wigner function. The density operator of the mixed state of $|\alpha\rangle$ and $|-\alpha\rangle$ is

$$\hat{\rho}_{\alpha\text{mix}} = \frac{1}{2}\Big(|\alpha\rangle\langle\alpha| + |-\alpha\rangle\langle-\alpha|\Big). \tag{1.139}$$

The Wigner function $W_{\alpha\text{mix}}(x, p)$ becomes therefore

$$
\begin{aligned}
W_{\alpha\text{mix}}(x, p) &= \frac{1}{\pi}\int_{-\infty}^{\infty} d\xi\, e^{-2ip\xi} \frac{1}{2}\Big[\langle x + \xi/2|\alpha\rangle\langle\alpha|x - \xi/2\rangle \\
&\qquad\qquad\qquad + \langle x + \xi/2|-\alpha\rangle\langle-\alpha|x - \xi/2\rangle\Big] \\
&= \frac{1}{\pi}\int_{-\infty}^{\infty} d\xi\, e^{-2ip\xi} \frac{1}{2}\Big[\langle x + \xi/2|\alpha\rangle\langle\alpha|x - \xi/2\rangle \\
&\qquad\qquad\qquad + \langle x + \xi/2|-\alpha\rangle\langle-\alpha|x - \xi/2\rangle\Big] \\
&= \frac{1}{2}W_\alpha(x, p) + \frac{1}{2}W_{-\alpha}(x, p) \\
&= \frac{1}{\pi}\Big[e^{-2(x-x_0)^2 - 2(p-p_0)^2} + e^{-2(x+x_0)^2 - 2(p+p_0)^2}\Big]. \tag{1.140}
\end{aligned}
$$

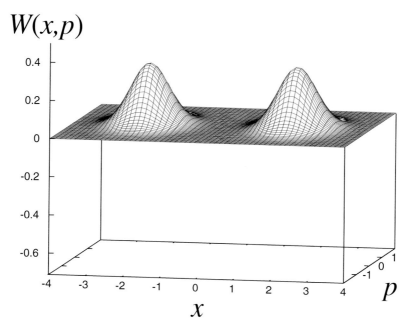

Fig. 1.36 The Wigner function of a mixed state of two coherent states $|\alpha\rangle$ and $|-\alpha\rangle$, where $(x_0, p_0) = (2, 0)$

Figure 1.36 shows the Wigner function of a mixed state of two coherent states $|\alpha\rangle$ and $|-\alpha\rangle$. The difference from a minus cat state (Fig. 1.35) is obvious. In the case of the mixed state there is no quantum interference structure at all and only two structures of coherent states $|\alpha\rangle$ and $|-\alpha\rangle$. So there is no negativity of the Wigner function and it is just a classical probability density.

So far we have explained minus cat states. We mentioned that the negativity of the Wigner function is a sign of nonclassicality. Now we will make a detour. We will show the Wigner functions of photon-number states (Fock states). That is because these are nonclassical states which have negativities in the Wigner function.

First of all, we will calculate the Wigner function of a single-photon state $|1\rangle$. Since the density operator of a single-photon state $|1\rangle$ is $\hat{\rho}_1 = |1\rangle\langle 1|$, we have to calculate $\psi_1(x) \equiv \langle x|1\rangle$ to calculate the Wigner function of a single-photon state $|1\rangle$. Fortunately we have already done so and got Eq. (1.96), also given in Eq. (1.105). It is

$$\langle x|1\rangle = 2\left(\frac{2}{\pi}\right)^{\frac{1}{4}} xe^{-x^2}. \tag{1.141}$$

By using this we can calculate the Wigner function of a single-photon state $|1\rangle$ as follows:

$$W_1(x, p) = \frac{1}{\pi} \int d\xi e^{-2ip\xi} \langle x + \xi/2 | 1 \rangle \langle 1 | x - \xi/2 \rangle$$

$$= \frac{1}{\pi} \int d\xi e^{-2ip\xi} \psi_1(x + \xi/2) \psi_1(x - \xi/2)$$

$$= \frac{1}{\pi} \int d\xi e^{-2ip\xi} 2 \left(\frac{2}{\pi}\right)^{\frac{1}{4}} (x + \xi/2) e^{-(x+\xi/2)^2}$$

$$\times 2 \left(\frac{2}{\pi}\right)^{\frac{1}{4}} (x - \xi/2) e^{-(x-\xi/2)^2}$$

$$= \frac{2}{\pi} e^{-2(x^2+p^2)} (4x^2 + 4p^2 - 1). \tag{1.142}$$

Figure 1.37 shows the Wigner function of a single-photon state $|1\rangle$. From the figure we can see that the Wigner function takes a negative value around the origin $(0, 0)$. So we can say that the single-photon state $|1\rangle$ is a nonclassical state and not a classical particle.

Similar to a single-photon state $|1\rangle$, we can calculate the Wigner functions of a two-photon state $|2\rangle$, a three-photon state $|3\rangle$, and a four-photon state $|4\rangle$ ($W_2(x, p)$, $W_3(x, p)$, and $W_4(x, p)$, respectively) with Eqs. (1.106), (1.107), and (1.108). They are the following:

$$W_2(x, p) = \frac{2}{\pi} e^{-2(x^2+p^2)} \left[8(x^2 + p^2)^2 - 8(x^2 + p^2) + 1\right], \tag{1.143}$$

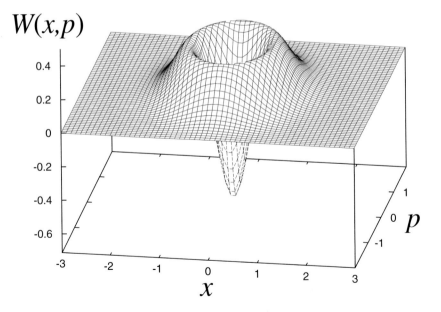

Fig. 1.37 The Wigner function of a single-photon state $|1\rangle$

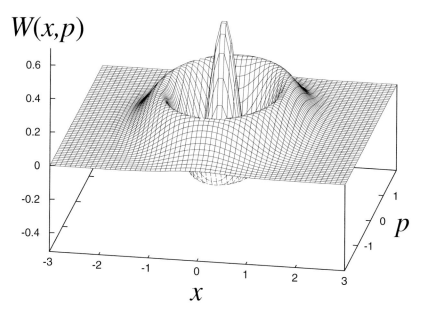

Fig. 1.38 The Wigner function of a two-photon state $|2\rangle$

$$W_3(x, p) = \frac{2}{\pi}e^{-2(x^2+p^2)}\left[\frac{32}{3}(x^2+p^2)^3 - 24(x^2+p^2)^2 + 12(x^2+p^2) - 1\right] \quad (1.144)$$

$$W_4(x, p) = \frac{2}{\pi}e^{-2(x^2+p^2)}\left[\frac{32}{3}(x^2+p^2)^4 - \frac{128}{3}(x^2+p^2)^3 + 48(x^2+p^2)^2\right.$$
$$\left. - 16(x^2+p^2) + 1\right]. \quad (1.145)$$

Figures 1.38, 1.39, and 1.40 show the Wigner functions of a two-photon state $|2\rangle$, a three-photon state $|3\rangle$, and a four-photon state $|4\rangle$ ($W_2(x, p)$, $W_3(x, p)$, and $W_4(x, p)$), respectively.

As mentioned before, there is a relation between a Wigner function and a marginal distribution (Eq. (1.138)). So we can understand the meaning of the Wigner functions by comparing Figs. 1.38 and 1.16, Figs. 1.39 and 1.17, and Figs. 1.40 and 1.18. For example, there is no phase dependence in Figs. 1.16, 1.17, and 1.18, which corresponds to that the respective Wigner functions have rotation symmetries around the vertical axis at the origin. Moreover, we can see zero-probability parts of amplitudes in the marginal distributions, which correspond to the negativities of the Wigner functions.

As a last note, we show some convenient things about the Wigner functions. One is that we can easily switch to another unit system, e.g., $\hbar = 1$ and $\hat{a} = (\hat{x} + i\hat{p})/\sqrt{2}$, from the present unit system where $\hbar = 1/2$ and $\hat{a} = \hat{x} + i\hat{p}$. That should be completed by just changing $\sqrt{2}x$ and $\sqrt{2}p$ to x and p, respectively. Of course, it is not a fundamental change but only a scale change. Another convenient thing is

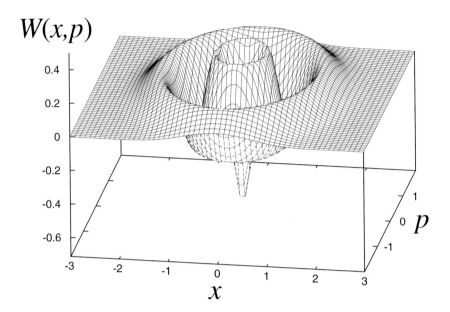

Fig. 1.39 The Wigner function of a three-photon state $|3\rangle$

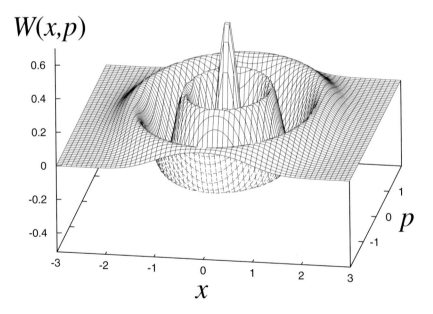

Fig. 1.40 The Wigner function of a four-photon state $|4\rangle$

that we have a formula for calculating the Wigner functions of photon-number states (Fock states) $W_n(x, p)$, that is,

$$W_n(x, p) = (-1)^n \frac{2}{\pi} e^{-2(x^2+p^2)} L_n(4x^2 + 4p^2), \quad (1.146)$$

where $L_n(\xi)$ is a Laguerre polynomial, which is

$$L_n(\xi) = \frac{e^\xi}{n!} \frac{d^n}{d\xi^n} (e^{-\xi} \xi^n). \quad (1.147)$$

1.9 Superposition States of a Vacuum and a Two-Photon State

As shown in the previous sections, photon-number states have no phase dependence. However, in Sect. 1.6, we showed that their superposition have phase dependence, where we considered a superposition of a vacuum $|0\rangle$ and a single-photon state $|1\rangle$. The superposition state has a sine-wave, i.e., "transverse wave" structure. In this section we will consider a "longitudinal wave" of an electro-magnetic field. The typical example should be a superposition, $(|0\rangle + |2\rangle)/\sqrt{2}$. Let's calculate its Wigner function, i.e., $W_{0+2}(x, p)$.

First we calculate the density operator of $(|0\rangle + |2\rangle)/\sqrt{2}$, $\hat{\rho}_{0+2}$:

$$\hat{\rho}_{0+2} = \left(\frac{1}{\sqrt{2}}\right)^2 (|0\rangle + |2\rangle)(\langle 0| - \langle 2|)$$

$$= \frac{1}{2} (|0\rangle\langle 0| + |2\rangle\langle 2| + |0\rangle\langle 2| + |2\rangle\langle 0|). \quad (1.148)$$

Then we can calculate the Wigner function $W_{0+2}(x, p)$:

$$W_{0+2}(x, p) = \frac{1}{\pi} \int_{-\infty}^{\infty} d\xi e^{-2ip\xi} \frac{1}{2} \Big[\langle x + \xi/2|0\rangle\langle 0|x - \xi/2\rangle$$

$$+ \langle x + \xi/2|2\rangle\langle 2|x - \xi/2\rangle$$

$$+ \langle x + \xi/2|0\rangle\langle 2|x - \xi/2\rangle$$

$$+ \langle x + \xi/2|2\rangle\langle 0|x - \xi/2\rangle \Big]$$

$$= \frac{1}{2} W_0(x, p) + \frac{1}{2} W_2(x, p)$$

$$+ \frac{1}{\pi} \int_{-\infty}^{\infty} d\xi e^{-2ip\xi} \frac{1}{2} \Big[\varphi_2 (x + \xi/2) \varphi_0 (x - \xi/2)$$

$$+ \varphi_0 (x + \xi/2) \varphi_2 (x - \xi/2) \Big]$$

$$= \frac{1}{2} W_0(x, p) + \frac{1}{2} W_2(x, p)$$

$$+ \frac{1}{\pi} \int_{-\infty}^{\infty} d\xi e^{-2ip\xi} \frac{1}{2} \left\{ \left(\frac{1}{2\pi} \right)^{\frac{1}{4}} \left[4(x + \xi/2)^2 - 1 \right]^2 e^{-(x+\xi/2)^2} \right.$$

$$\times \left(\frac{2}{\pi} \right)^{\frac{1}{4}} e^{-(x-\xi/2)^2}$$

$$+ \left(\frac{1}{2\pi} \right)^{\frac{1}{4}} \left[4(x - \xi/2)^2 - 1 \right]^2 e^{-(x-\xi/2)^2}$$

$$\left. \times \left(\frac{2}{\pi} \right)^{\frac{1}{4}} e^{-(x+\xi/2)^2} \right\}$$

$$= \frac{1}{2} \cdot \frac{2}{\pi} e^{-2(x^2+p^2)}$$

$$+ \frac{1}{2} \cdot \frac{2}{\pi} e^{-2(x^2+p^2)} \left[8(x^2 + p^2)^2 - 8(x^2 + p^2) + 1 \right]$$

$$+ \frac{4\sqrt{2}}{\pi} e^{-2(x^2+p^2)} (x^2 - p^2)$$

$$= \frac{2}{\pi} e^{-2(x^2+p^2)} \left[4(x^2 + p^2)^2 - 4(x^2 + p^2) + 2\sqrt{2}(x^2 - p^2) + 1 \right], \tag{1.149}$$

where we used Eqs. (1.104) and (1.106).

Figure 1.41 shows $W_{0+2}(x, p)$ and Fig. 1.42 shows the view in the x direction. Moreover, we calculate the marginal distributions with Eq. (1.138) from the Wigner function and get the phase dependence of amplitude of $(|0\rangle + |2\rangle)/\sqrt{2}$, which is shown in Fig. 1.43.

It is obvious from Fig. 1.43 that the phase dependence of amplitude of $(|0\rangle + |2\rangle)/\sqrt{2}$ has no sine-wave, i.e., "transverse wave" structure. Instead, we can see a "compressional wave" or "longitudinal wave" structure. This structure can also be seen in the Wigner function shown in Fig. 1.42, in which there is dyad symmetry and the rotation corresponds to time evolution. In the case of no rotation symmetry in the Wigner function, the phase dependence of amplitude has to have a "transverse wave" structure. That is because the Wigner function comes back to the original one after one rotation, which creates a transverse-wave structure. However, in the case of dyad symmetry like in the Wigner function of $(|0\rangle + |2\rangle)/\sqrt{2}$, it creates a "compressional-wave" or "longitudinal-wave" structure. In that sense, since a minus cat state has dyad symmetry in the Wigner function as shown in Figs. 1.34 and 1.35, it also can be regarded as a "compressional wave" or "longitudinal wave" of an electromagnetic field. Figure 1.44 shows the Wigner function of superposition of a vacuum $|0\rangle$ and a single-photon state $|1\rangle$, $(|0\rangle + |1\rangle)/\sqrt{2}$. From the figure we do not see any rotation symmetry. So the phase dependence of amplitude have a sine-wave structure as shown in Fig. 1.22. Note that the frequency of a wave structure without rotation

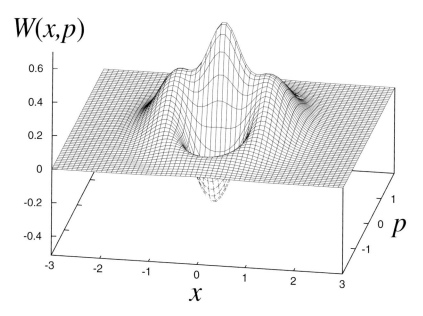

Fig. 1.41 The Wigner function of a superposition of a vacuum $|0\rangle$ and a two-photon state $|2\rangle$, $(|0\rangle + |2\rangle)/\sqrt{2}$

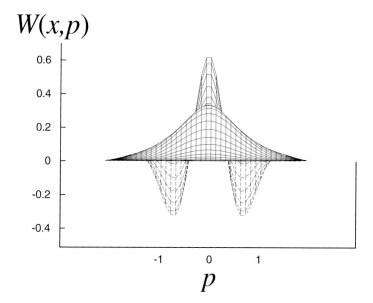

Fig. 1.42 View of the Wigner function $W_{0+2}(x, p)$ from the x direction

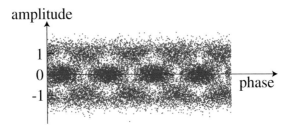

Fig. 1.43 Phase dependence of amplitude of $(|0\rangle + |2\rangle)/\sqrt{2}$

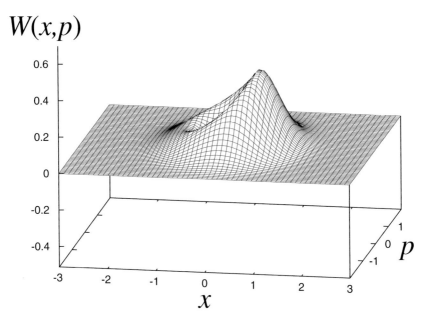

Fig. 1.44 The Wigner function of a superposition of a vacuum $|0\rangle$ and a single-photon state $|1\rangle$, $(|0\rangle + |1\rangle)/\sqrt{2}$. There is no rotation symmetry

symmetry in the Wigner function is the same as the one of the electro-magnetic field, but it becomes twice as large when the Wigner function shows dyad symmetry.

1.10 Squeezed States

In Sect. 1.7 we considered a wave structure of a coherent state as the extension of that of $(|0\rangle + |1\rangle)/\sqrt{2}$. In this section we will consider a "compressional-wave" or "longitudinal-wave" structure of a squeezed vacuum as the extension of that of $(|0\rangle + |2\rangle)/\sqrt{2}$.

A squeezed vacuum is defined by using photon-number states:

$$\hat{S}(\xi)|0\rangle = (1 - |\kappa|^2)^{\frac{1}{4}} \sum_{n=0}^{\infty} \frac{\sqrt{(2n)!}}{2^n n!} \kappa^n |2n\rangle. \qquad (1.150)$$

Here $\xi = |\xi|e^{i\varphi\xi}$, $\kappa = e^{i\varphi\xi}\tanh|\xi|$, ξ is a squeezing parameter, φ is a squeezing phase, and $\hat{S}(\xi)$ is a squeezing operator. Of course, $|0\rangle$ is a vacuum. We now define r as the value of ξ at $\varphi = 0$ and treat it as a squeezing parameter hereafter. We can do so without loss of generality. Then Eq. (1.150) becomes

$$\hat{S}(r)|0\rangle = (1 - \tanh^2 r)^{\frac{1}{4}} \sum_{n=0}^{\infty} \frac{\sqrt{(2n)!}}{2^n n!} \tanh^n r |2n\rangle$$

$$= (1 - \tanh^2 r)^{\frac{1}{4}} \left(\frac{\sqrt{0!}}{2^0 0!} \tanh^0 r |0\rangle + \frac{\sqrt{2!}}{2^1 1!} \tanh^1 r |2\rangle \right.$$

$$\left. + \frac{\sqrt{4!}}{2^2 2!} \tanh^2 r |4\rangle + \frac{\sqrt{6!}}{2^3 3!} \tanh^3 r |6\rangle + \cdots \right). \quad (1.151)$$

We consider $-3\,$dB(decibel)-squeezing, which is experimentally feasible. When the squeezing level is 3 dB, we can get $\tanh r = 1/3$ by using $e^{-2r} = 0.5$ ($r = \ln 2/2$). Then the $-3\,$dB-squeezed vacuum can be written as

$$\hat{S}(\ln 2/2)|0\rangle = \left(1 - \frac{1}{3^2} \right)^{\frac{1}{4}} \left(|0\rangle + \frac{1}{3\sqrt{2}}|2\rangle \cdots \right)$$

$$= 0.971|0\rangle + 0.229|2\rangle + \cdots. \qquad (1.152)$$

Here we have neglected states with photon number higher than two. It can be justified by the fact that $0.971^2 + 0.229^2 = 0.995$. So the $-3\,$dB-squeezed vacuum can be well approximated by a superposition of a vacuum $|0\rangle$ and a two-photon state $|2\rangle$. The Wigner function $W_{3\,\text{dB-Sqz}}(x, p)$ can be easily calculated with Eq. (1.149) as follows:

$$W_{3\,\text{dB-Sqz}}(x, p) = \frac{1}{\pi} \int_{-\infty}^{\infty} d\xi e^{-2ip\xi} \Big[0.971^2 \langle x + \xi/2|0\rangle \langle 0|x - \xi/2\rangle$$

$$+ 0.229^2 \langle x + \xi/2|2\rangle \langle 2|x - \xi/2\rangle$$

$$+ 0.971 \times 0.229 \langle x + \xi/2|0\rangle \langle 2|x - \xi/2\rangle$$

$$+ 0.971 \times 0.229 \langle x + \xi/2|2\rangle \langle 0|x - \xi/2\rangle \Big]$$

$$= 0.943 W_0(x, p) + 0.0524 W_2(x, p)$$

$$+ \frac{1}{\pi} \int_{-\infty}^{\infty} d\xi e^{-2ip\xi} 0.222 \Big[\varphi_2(x + \xi/2)\, \varphi_0(x - \xi/2)$$

$$+ \varphi_0(x + \xi/2)\, \varphi_2(x - \xi/2) \Big]$$

$$= 0.943 \cdot \frac{2}{\pi} e^{-2(x^2 + p^2)}$$

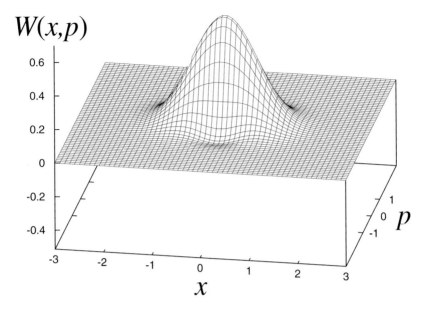

Fig. 1.45 The Wigner function of a 3 dB-squeezed vacuum approximated by a superposition of a vacuum $|0\rangle$ and a two-photon state $|2\rangle$. Although there is some negativity in the Wigner function, it is an artifact caused by lack of accuracy in the approximation due to omission of higher order terms. If we take more photons into account, the negativity should vanish

$$+ 0.0524 \cdot \frac{2}{\pi} e^{-2(x^2+p^2)} \left[8(x^2 + p^2)^2 - 8(x^2 + p^2) + 1 \right]$$

$$+ 0.222 \cdot \frac{8\sqrt{2}}{\pi} e^{-2(x^2+p^2)} (x^2 - p^2)$$

$$= \frac{2}{\pi} e^{-2(x^2+p^2)} \Big[0.419(x^2 + p^2)^2 - 0.419(x^2 + p^2)$$

$$+ 1.26(x^2 - p^2) + 0.995 \Big]. \tag{1.153}$$

Figure 1.45 shows the Wigner function $W_{3\,\text{dB-Sqz}}(x, p)$. The view in the x direction is shown in Fig. 1.46. Although there is some negativity in the Wigner function, it is an artifact caused by lack of accuracy in the approximation due to omission of higher order terms. If we take more photons into account, the negativity should vanish.

We can see that the Wigner function has a dyad symmetry as shown in Fig. 1.45. So the state should show a "compressional-wave" or "longitudinal-wave" structure in the phase dependence of amplitude. Figure 1.47 shows the phase dependence of amplitude. We can see a "compressional-wave" or "longitudinal-wave" structure there. Moreover, we can see the reason why the state is called a "squeezed" state, that is, the Wigner function of a vacuum $|0\rangle$ (Fig. 1.48) is "squeezed" in the p direction and get the shape of Fig. 1.45. It is "anti-squeezed" in the x direction. Therefore

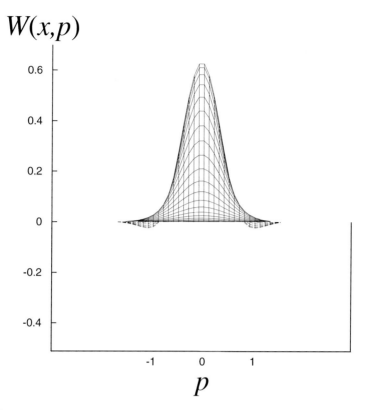

Fig. 1.46 The view of Fig. 1.45 in the x direction

Fig. 1.47 Phase dependence
of amplitude of a
3 dB-squeezed vacuum
approximated by a
superposition of a vacuum
$|0\rangle$ and a two-photon state
$|2\rangle$

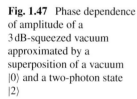

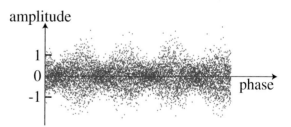

the variance of amplitude becomes small in one quadrature component (p) and it
becomes large in the other quadrature component (x) in Fig. 1.47. Here the minimum
variance is smaller than the one of a vacuum in this case. So a squeezed vacuum can
be regarded as a nonclassical state in this sense and we sometimes call only this case
squeezing.

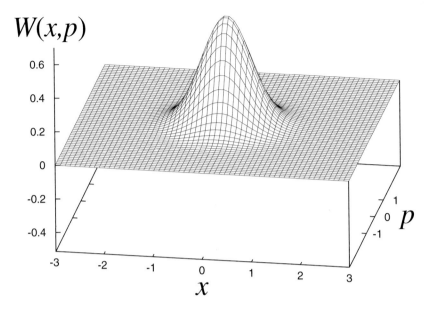

Fig. 1.48 The Wigner function of a vacuum $|0\rangle$

1.11   Squeezing Operation and Squeezed States

We will describe a squeezed vacuum and general squeezed states in a different way from the previous section in order to explain them more extensively. We will do it in historical order.

A coherent state is an eigenstate of annihilation operator $\hat{a}$ as mentioned in Sect. 1.2. We make a Bogoliubov transformation of operator $\hat{a}$ and get operator $\hat{b}$ as follows:

$$\hat{b} = \mu\hat{a} + \nu\hat{a}^{\dagger} \quad (|\mu|^2 - |\nu|^2 = 1). \tag{1.154}$$

Squeezed states will then be the eigenstates of operator $\hat{b}$. Here the Bogoliubov transformation is called a squeezing operation. Note that the Bogoliubov transformation was introduced to explain the process of creation of Cooper pairs of electrons in superconducting materials in the BCS theory. Therefore the squeezing process makes pairwise photons, where a photon is a Boson and an electron is a Fermion and we have to be careful about the difference between the commutation relationship and the anti-commutation relationship.

The Bogoliubov transformation can be regarded as time-evolution of an operator in the Heisenberg picture. In the Schrödinger picture we treat time-evolution of a quantum state $|\psi\rangle$ with the Schrödinger equation (Eq. (1.41)) as follows:

$$|\psi(t)\rangle = e^{-i\frac{\hat{H}}{\hbar}t}|\psi(0)\rangle. \tag{1.155}$$

In the Heisenberg picture we treat time-evolution of an operator $\hat{A}$ with the Heisenberg equation of motion (Eq. (1.8)) as follows:

$$\hat{A}(t) = e^{i\frac{\hat{H}}{\hbar}t}\hat{A}(0)e^{-i\frac{\hat{H}}{\hbar}t}. \tag{1.156}$$

Here we define a squeezing operator $\hat{S}(r)$ as

$$\hat{S}(r) = e^{\frac{r}{2}(\hat{a}^{\dagger 2} - \hat{a}^2)}. \tag{1.157}$$

Note that $r \in \mathcal{R}$ is the squeezing parameter ξ at $\varphi = 0$ as mentioned in the previous section.

When we set the Hamiltonian of squeezing as

$$\hat{H}_{\text{sqz}} = \frac{i}{2}\left(\hat{a}^2 - \hat{a}^{\dagger 2}\right), \tag{1.158}$$

the squeezing operator $S(r)$ can be regarded as the time-evolution operator $e^{-i\frac{\hat{H}_{\text{sqz}}}{\hbar}t}$ with $r \to t$. So the squeezing operation can be regarded as the time-evolution of quantum complex amplitude $\hat{a}$ with Eq. (1.156). Thus it can be written as follows:

$$\begin{aligned}
\hat{a}(t) &= e^{i\frac{\hat{H}_{\text{sqz}}}{\hbar}t}\hat{a}e^{-i\frac{\hat{H}_{\text{sqz}}}{\hbar}t} \\
&= \hat{S}^{\dagger}(r)\hat{a}\hat{S}(r) \\
&= e^{-\frac{r}{2}(\hat{a}^{\dagger 2} - \hat{a}^2)}\hat{a}e^{\frac{r}{2}(\hat{a}^{\dagger 2} - \hat{a}^2)} \\
&= \hat{a} + \left[-\frac{r}{2}(\hat{a}^{\dagger 2} - \hat{a}^2), \hat{a}\right] \\
&\quad + \frac{1}{2!}\left[-\frac{r}{2}(\hat{a}^{\dagger 2} - \hat{a}^2), \left[-\frac{r}{2}(\hat{a}^{\dagger 2} - \hat{a}^2), \hat{a}\right]\right] + \cdots \\
&= \hat{a} + r\hat{a}^{\dagger} + \frac{1}{2!}\left[-\frac{r}{2}(\hat{a}^{\dagger 2} - \hat{a}^2), r\hat{a}\right] + \cdots \\
&= \hat{a} + r\hat{a}^{\dagger} + \frac{r^2}{2!}\hat{a} + \frac{r^3}{3!}\hat{a}^{\dagger} + \cdots \\
&= \hat{a}\cosh r + \hat{a}^{\dagger}\sinh r. \tag{1.159}
\end{aligned}$$

Here we used the formula

$$e^{\zeta\hat{B}}\hat{A}e^{-\zeta\hat{B}} = \hat{A} + \zeta[\hat{B}, \hat{A}] + \frac{\zeta^2}{2!}[\hat{B}, [\hat{B}, \hat{A}]] + \cdots . \tag{1.160}$$

It is obvious that the time-evolution corresponds to the Bogoliubov transformation with $\cosh^2 r - \sinh^2 r = 1$. Moreover, we can calculate a squeezed vacuum with Eq. (1.157) and

$$\hat{S}(r)|0\rangle = e^{\frac{r}{2}(\hat{a}^{\dagger 2} - \hat{a}^2)}|0\rangle, \tag{1.161}$$

and get Eq. (1.151). Of course, we can derive Eq. (1.150) with $r \to \xi$. In any case, since the created state is a superposition of even-photon-number states, it follows that photon-photon correlation is created through the Bogoliubov transformation.

I think it is not clear why the eigenstates of $\hat{b}$ is called squeezed states from the explanation above. We will explain the reason here. It should be obvious if we see the following equation:

$$\hat{b} = \hat{a} \cosh r + \hat{a}^\dagger \sinh r = \hat{x}e^r + i\hat{p}e^{-r}. \tag{1.162}$$

It is extended in x-direction and shrunk in p-direction. Of course, it can be seen as the time-evolution as follows:

$$\hat{S}^\dagger(r)\hat{x}\hat{S}(r) = \hat{S}^\dagger(r)\left(\frac{\hat{a}+\hat{a}^\dagger}{2}\right)\hat{S}(r)$$

$$= \frac{1}{2}\left(\hat{a}\cosh r + \hat{a}^\dagger \sinh r + \hat{a}^\dagger \cosh r + \hat{a}\sinh r\right)$$

$$= \hat{x}e^r, \tag{1.163}$$

and

$$\hat{S}^\dagger(r)\hat{p}\hat{S}(r) = \hat{S}^\dagger(r)\left(\frac{\hat{a}-\hat{a}^\dagger}{2i}\right)\hat{S}(r)$$

$$= \frac{1}{2i}\left(\hat{a}\cosh r + \hat{a}^\dagger \sinh r - \hat{a}^\dagger \cosh r - \hat{a}\sinh r\right)$$

$$= \hat{p}e^{-r}. \tag{1.164}$$

However, they are in any case just transformations of operators and we do not get much information on the transformation from these equations. So we will calculate the average values and variances of the corresponding physical quantities with the operators.

First we will calculate the average values of $\hat{x}$, $\hat{p}$, $\hat{x}^2$, and $\hat{p}^2$, respectively:

$$\langle 0|\hat{S}^\dagger(r)\hat{x}\hat{S}(r)|0\rangle = \langle 0|\hat{S}^\dagger(r)\left(\frac{\hat{a}+\hat{a}^\dagger}{2}\right)\hat{S}(r)|0\rangle$$

$$= \langle 0|\frac{1}{2}\left(\hat{a}\cosh r + \hat{a}^\dagger \sinh r + \hat{a}^\dagger \cosh r + \hat{a}\sinh r\right)|0\rangle$$

$$= 0, \tag{1.165}$$

$$\langle 0|\hat{S}^\dagger(r)\hat{p}\hat{S}(r)|0\rangle = 0, \tag{1.166}$$

$$\langle 0|\hat{S}^\dagger(r)\hat{x}^2\hat{S}(r)|0\rangle = \frac{1}{4}e^{2r}, \tag{1.167}$$

Fig. 1.49 Image of a squeezed vacuum in phase space

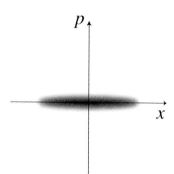

$$\langle 0|\hat{S}^{\dagger}(r)\hat{p}^2\hat{S}(r)|0\rangle = \frac{1}{4}e^{-2r}. \tag{1.168}$$

Then we can calculate the variances, i.e., quantum fluctuation, with Eq. (1.18) and get

$$\Delta x_{\text{sqz}} = \frac{1}{2}e^{r}, \tag{1.169}$$

and

$$\Delta p_{\text{sqz}} = \frac{1}{2}e^{-r}. \tag{1.170}$$

From these results we can see the average values of x and p components are both zero, which are the same as the ones of a vacuum $|0\rangle$. The quantum fluctuations of x and p components, however, are asymmetrical. Furthermore, we can see that the state is a minimum-uncertainty state from $\Delta x_{\text{sq}} \Delta p_{\text{sq}} = \frac{1}{4}$. Thus the image of a squeezed vacuum is shown in Fig. 1.49. As is clear from the figure, the name "squeeze" comes from the asymmetry of quantum fluctuation between x and p components. Note that we can "rotate" the asymmetry with $r \to \xi$, which is defined in Eq. (1.150).

Now we calculate the Wigner function of a squeezed vacuum. For that purpose we introduce the following equation:

$$\hat{S}(r)|x\rangle = |xe^{r}\rangle. \tag{1.171}$$

This relation can be checked in the following way:

$$\hat{x}|x\rangle = x|x\rangle,$$
$$\hat{S}(r)\hat{x}|x\rangle = \hat{S}(r)x|x\rangle,$$
$$\hat{S}(r)\hat{x}\hat{S}^{\dagger}(r)\hat{S}(r)|x\rangle = \hat{S}(r)x|x\rangle,$$
$$\hat{x}e^{-r}\hat{S}(r)|x\rangle = \hat{S}(r)x|x\rangle,$$
$$\hat{x}(\hat{S}(r)|x\rangle) = xe^{r}(\hat{S}(r)|x\rangle). \tag{1.172}$$

Since the last equation shows that $\hat{S}(r)|x\rangle$ is an eigenfunction of operator $\hat{x}$ with the eigenvalue of xe^r, we can check Eq. (1.171). Here we use Eq. (1.163) and $\hat{S}^\dagger(r)\hat{S}(r) = \hat{1}$.

So the Wigner function of a squeezed vacuum W_{0sqz} can be calculated as follows:

$$
\begin{aligned}
W_{0sqz}(x, p) &= \frac{1}{\pi} \int d\xi e^{-2ip\xi} \langle x + \xi/2|\hat{S}(r)|0\rangle \langle 0|\hat{S}^\dagger(r)|x - \xi/2\rangle \\
&= \frac{e^{-r}}{\pi} \int d\xi e^{-2ip\xi} \langle (x + \xi/2)e^{-r}|0\rangle \langle 0|(x - \xi/2)e^{-r}\rangle \\
&= \frac{e^{-r}}{\pi} \int d\xi e^{-2ip\xi} \psi_0\left((x + \xi/2)e^{-r}\right) \psi_0\left((x - \xi/2)e^{-r}\right) \\
&= \frac{e^{-r}}{\pi} \int d\xi e^{-2ip\xi} \left(\frac{2}{\pi}\right)^{\frac{1}{4}} e^{-(x+\xi/2)^2 e^{-2r}} \left(\frac{2}{\pi}\right)^{\frac{1}{4}} e^{-(x-\xi/2)^2 e^{-2r}} \\
&= \frac{2}{\pi} e^{-2(x^2 e^{-2r} + p^2 e^{2r})}.
\end{aligned}
\tag{1.173}
$$

By using Eq. (1.173) we can get the Wigner function of a $-10\,$dB-squeezed vacuum $\hat{S}(r = 1.15)|0\rangle$ as shown in Fig. 1.50. From the Wigner function we can get the phase dependence of amplitude as shown in Fig. 1.51. We do not see any negativity in the Wigner function here in contrast to the case of approximation of a

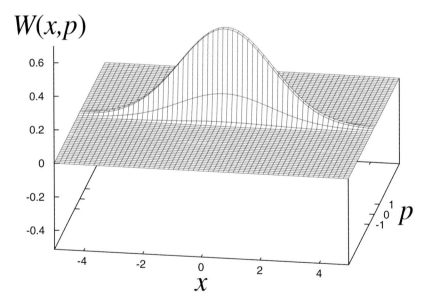

Fig. 1.50 The Wigner function of a $10\,$dB-squeezed vacuum $\hat{S}(r = 1.15)|0\rangle$. When $r = 1.15$, $e^{-2r} = 0.1$ and then the quantum fluctuations of the p component become $1/10 = -10\,$dB compared to the case of a vacuum

Fig. 1.51 Phase dependence of amplitude of a 10 dB-squeezed vacuum $\hat{S}(r = 1.15)|0\rangle$

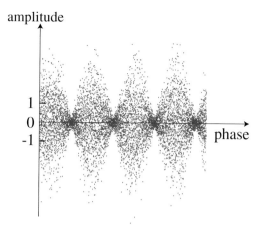

$-3\,$dB-squeezed vacuum $W_{3\,\text{dB-Sqz}}(x, p)$ with a superposition of a vacuum $|0\rangle$ and a two-photon state $|2\rangle$, which is shown in Fig. 1.45. It is because the Wigner function presented here is not an approximated one but a true one. Moreover, we can see a "longitudinal-wave" structure in Fig. 1.51.

1.12 Quantum Entanglement

So far we have dealt with one optical beam. We will now deal with two optical beams. Of course, it is extendable to more than two beams, but we only deal with two beams here for simplicity.

First we deal with the simplest case, where a single photon $|1\rangle$ enters a 50:50 beam splitter as shown in Fig. 1.52. Note that a vacuum $|0\rangle$ enters another port of the beam splitter. The output state of the two beams from the beam splitter $|\psi_{\text{ABent}}\rangle$ is described as

$$|\psi_{\text{ABent}}\rangle = \frac{1}{\sqrt{2}}\big(|1\rangle_A|0\rangle_B - |0\rangle_A|1\rangle_B\big). \qquad (1.174)$$

Fig. 1.52 A single photon $|1\rangle$ enters a 50:50 beam splitter

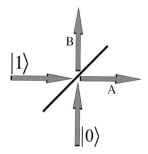

When a photon exists in beam A in this state then no photon exists in beam B and vice versa. When we make a measurement on one of the two outputs then the state of the other beam gets determined. The two beams A and B are entangled. More precisely, the overall state of the two outputs A and B $|\psi_{ABent}\rangle$ cannot be described by a tensor product of the state of beam A $|\phi_A\rangle$ and that of beam B $|\varphi_B\rangle$;

$$|\psi_{ABent}\rangle \neq |\phi_A\rangle \otimes |\varphi_B\rangle. \tag{1.175}$$

Obviously $|\psi_{ABent}\rangle$ satisfies this relation.

Note that this entanglement cannot be distinguished with photon-number measurement from a classical correlation $\hat{\rho}_{ABclas}$:

$$\hat{\rho}_{ABclas} = \frac{1}{2}\Big(|1\rangle_A\langle 1| \otimes |0\rangle_B\langle 0| + |0\rangle_A\langle 0| \otimes |1\rangle_B\langle 1|\Big). \tag{1.176}$$

Even in this state when a photon exists in beam A then no photon exists in beam B and vice versa.

The difference between the entangled state $|\psi_{ABent}\rangle$ and the classically correlated state $\hat{\rho}_{ABclas}$ is the existence or non-existence of a non-local superposition.

Let's check this difference. We define $|+\rangle$ and $|-\rangle$ as

$$|+\rangle = \frac{|0\rangle + |1\rangle}{\sqrt{2}}, \tag{1.177}$$

$$|-\rangle = \frac{|0\rangle - |1\rangle}{\sqrt{2}}. \tag{1.178}$$

Then we can get the equations

$$|0\rangle = \frac{|+\rangle + |-\rangle}{\sqrt{2}}, \tag{1.179}$$

$$|1\rangle = \frac{|+\rangle - |-\rangle}{\sqrt{2}}. \tag{1.180}$$

By using above equations we deform Eq. (1.174) and get

$$\frac{1}{\sqrt{2}}\big(|1\rangle_A|0\rangle_B - |0\rangle_A|1\rangle_B\big) = \frac{1}{\sqrt{2}}\Bigg(\frac{|+\rangle_A - |-\rangle_A}{\sqrt{2}} \otimes \frac{|+\rangle_B + |-\rangle_B}{\sqrt{2}}$$
$$-\frac{|+\rangle_A + |-\rangle_A}{\sqrt{2}} \otimes \frac{|+\rangle_B - |-\rangle_B}{\sqrt{2}}\Bigg)$$
$$= \frac{1}{\sqrt{2}}\big(|+\rangle_A|-\rangle_B - |-\rangle_A|+\rangle_B\big). \tag{1.181}$$

In this equation we can see that when the state of beam A is $|+\rangle$ then the state of beam B is $|-\rangle$ and when the state of beam A is $|-\rangle$ then the state of beam B is $|+\rangle$.

There is still a correlation between beams A and B. Since the transformation from
$|0\rangle$ and $|1\rangle$ to $|+\rangle$ and $|-\rangle$ corresponds to basis transformation, it follows that the
entanglement is preserved under basis transformation.

On the other hand, when we make the same basis transformation for the classically
correlated state of Eq. (1.176), we get the following equation:

$$
\frac{1}{2}\Big(|1\rangle_A\langle 1| \otimes |0\rangle_B\langle 0| + |0\rangle_A\langle 0| \otimes |1\rangle_B\langle 1|\Big)
$$

$$
= \frac{1}{2}\Big(\frac{|+\rangle_A - |-\rangle_A}{\sqrt{2}} \frac{\langle+|_A - \langle-|_A}{\sqrt{2}} \otimes \frac{|+\rangle_B + |-\rangle_B}{\sqrt{2}} \frac{\langle+|_B + \langle-|_B}{\sqrt{2}}
$$

$$
+ \frac{|+\rangle_A + |-\rangle_A}{\sqrt{2}} \frac{\langle+|_A + \langle-|_A}{\sqrt{2}} \otimes \frac{|+\rangle_B - |-\rangle_B}{\sqrt{2}} \frac{\langle+|_B - \langle-|_B}{\sqrt{2}}\Big)
$$

$$
= \frac{1}{8}\Big[\Big(|+\rangle_A\langle+| + |-\rangle_A\langle-| - |+\rangle_A\langle-| - |-\rangle_A\langle+|\Big)
$$

$$
\otimes \Big(|+\rangle_B\langle+| + |-\rangle_B\langle-| + |+\rangle_B\langle-| + |-\rangle_B\langle+|\Big)
$$

$$
+ \Big(|+\rangle_A\langle+| + |-\rangle_A\langle-| + |+\rangle_A\langle-| + |-\rangle_A\langle+|\Big)
$$

$$
\otimes \Big(|+\rangle_B\langle+| + |-\rangle_B\langle-| - |+\rangle_B\langle-| - |-\rangle_B\langle+|\Big)\Big]
$$

$$
= \frac{1}{4}\Big[\Big(|+\rangle_A\langle+| + |-\rangle_A\langle-|\Big) \otimes \Big(|+\rangle_B\langle+| + |-\rangle_B\langle-|\Big)
$$

$$
- \Big(|+\rangle_A\langle-| + |-\rangle_A\langle+|\Big) \otimes \Big(|+\rangle_B\langle-| + |-\rangle_B\langle+|\Big)\Big]. \tag{1.182}
$$

From the result we can see that beams A and B have no correlation in this case.
The original correlation in the classically correlated state $\hat{\rho}_{ABclas}$ vanishes when we
change the measurement basis. So it follows that in the case of entangled states
a correlation exists even after changing the measurement basis but in the case of
classically correlated states it doesn't.

Where does the entanglement of the state $|\psi_{ABent}\rangle$ come from? It is obvious when
we look at the density operator of state $|\psi_{ABent}\rangle$:

$$
\hat{\rho}_{ABent} = \frac{|1\rangle_A|0\rangle_B - |0\rangle_A|1\rangle_B}{\sqrt{2}} \frac{\langle 1|_A\langle 0|_B - \langle 0|_A\langle 1|_B}{\sqrt{2}}
$$

$$
= \frac{1}{2}\Big(|1\rangle_A\langle 1| \otimes |0\rangle_B\langle 0| + |0\rangle_A\langle 0| \otimes |1\rangle_B\langle 1|
$$

$$
- |1\rangle_A\langle 0| \otimes |0\rangle_B\langle 1| - |0\rangle_A\langle 1| \otimes |1\rangle_B\langle 0|\Big)
$$

$$
= \hat{\rho}_{ABclas}
$$

$$
- \frac{1}{2}\Big(|1\rangle_A\langle 0| \otimes |0\rangle_B\langle 1| + |0\rangle_A\langle 1| \otimes |1\rangle_B\langle 0|\Big). \tag{1.183}
$$

Fig. 1.53 Photon-number
states $|n_1\rangle$ and $|n_2\rangle$ enter a
beam splitter

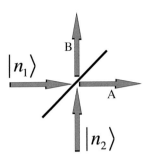

In this density operator there is off-diagonal terms on top of the classical correlation term $\hat{\rho}_{\text{ABclas}}$. In other words, the entanglement is preserved even after changing the measurement basis like in Eq. (1.181) because of the existence of the off-diagonal terms. Off-diagonal terms make entanglement.

So far we have discussed how to create entanglement with a single photon $|1\rangle$ and a 50:50 beam splitter. We can extend this methodology to arbitrary photon-number states and beam splitters. Let's think about the situation shown in Fig. 1.53, where photon-number states $|n_1\rangle$ and $|n_2\rangle$ enter a beam splitter. The input state can be described as $|n_1\rangle_A \otimes |n_2\rangle_B$ but we describe it as $|n_1, n_2\rangle$ for simplicity. The output state can be described with a beam splitter operator $\hat{B}$, which was introduced in Sect. 1.3.1, as $\hat{B}|n_1, n_2\rangle$. We calculate the output state as follows:

$$\hat{B}|n_1, n_2\rangle = \hat{B}\frac{1}{\sqrt{n_1! n_2!}}\hat{a}_1^{\dagger n_1}\hat{a}_2^{\dagger n_2}|0, 0\rangle$$

$$= \hat{B}\frac{1}{\sqrt{n_1! n_2!}}\hat{a}_1^{\dagger n_1}\hat{a}_2^{\dagger n_2}\hat{B}^{\dagger}|0, 0\rangle$$

$$= \frac{1}{\sqrt{n_1! n_2!}}\hat{B}\hat{a}_1^{\dagger n_1}\hat{B}^{\dagger}\hat{B}\hat{a}_2^{\dagger n_2}\hat{B}^{\dagger}|0, 0\rangle. \qquad (1.184)$$

Here the second equality comes from energy conservation $|0, 0\rangle = \hat{B}^{\dagger}|0, 0\rangle$, and the third equality comes from $\hat{B}^{\dagger}\hat{B} = \hat{I}$. Moreover, if we use $\hat{B}^{\dagger}\hat{B} = \hat{I}$, then we get the following equality for $\hat{B}\hat{a}_1^{\dagger n_1}\hat{B}^{\dagger}\hat{B}\hat{a}_2^{\dagger n_2}\hat{B}^{\dagger}$:

$$\hat{B}\hat{a}_1^{\dagger n_1}\hat{B}^{\dagger}\hat{B}\hat{a}_2^{\dagger n_2}\hat{B}^{\dagger} = \hat{B}\hat{a}_1\hat{B}^{\dagger} \cdot \hat{B}\hat{a}_1\hat{B}^{\dagger} \cdots \hat{B}\hat{a}_1\hat{B}^{\dagger} \cdot \hat{B}\hat{a}_2^{\dagger}\hat{B}^{\dagger} \cdot \hat{B}\hat{a}_2^{\dagger}\hat{B}^{\dagger} \cdots \hat{B}\hat{a}_2^{\dagger}\hat{B}^{\dagger}$$

$$= (B_{11}\hat{a}_1^{\dagger} + B_{21}\hat{a}_2^{\dagger})^{n_1}(B_{12}\hat{a}_1^{\dagger} + B_{22}\hat{a}_2^{\dagger})^{n_2}. \qquad (1.185)$$

Since the derivation of the second equality is a bit complicated, we will explain it in detail.

As explained in Sect. 1.3.1, a beam-splitter input-output relation can be described with a 2×2 matrix $\underline{B}$ and when two input field operators (annihilation operators) are $\hat{a}_1$ and $\hat{a}_2$ and two output field operators (annihilation operators) are $\hat{a}_1'$ and $\hat{a}_2'$, then the input-output relation is described as follows (Eq. (1.44)):

$$\begin{pmatrix} \hat{a}'_1 \\ \hat{a}'_2 \end{pmatrix} = \underline{B} \begin{pmatrix} \hat{a}_1 \\ \hat{a}_2 \end{pmatrix} = \begin{pmatrix} B_{11} & B_{12} \\ B_{21} & B_{22} \end{pmatrix} \begin{pmatrix} \hat{a}_1 \\ \hat{a}_2 \end{pmatrix}. \tag{1.186}$$

Since $\underline{B}$ is a unitary matrix, we have

$$\underline{B}^\dagger \begin{pmatrix} \hat{a}_1 \\ \hat{a}_2 \end{pmatrix} = \begin{pmatrix} B_{11}^* & B_{21}^* \\ B_{12}^* & B_{22}^* \end{pmatrix} \begin{pmatrix} \hat{a}'_1 \\ \hat{a}'_2 \end{pmatrix}. \tag{1.187}$$

We can describe the input-output relation with the beam splitter operator $\hat{B}$, that is,

$$\begin{pmatrix} \hat{a}'_1 \\ \hat{a}'_2 \end{pmatrix} = \hat{B}^\dagger \begin{pmatrix} \hat{a}_1 \\ \hat{a}_2 \end{pmatrix} \hat{B}. \tag{1.188}$$

Therefore, when we exchange the input and the output, i.e., input beams enter from the opposite side of the beam splitter, we have the following equation:

$$\begin{pmatrix} \hat{a}_1 \\ \hat{a}_2 \end{pmatrix} = \hat{B} \begin{pmatrix} \hat{a}'_1 \\ \hat{a}'_2 \end{pmatrix} \hat{B}^\dagger. \tag{1.189}$$

From Eqs. (1.187) and (1.189), we can get

$$\hat{B} \begin{pmatrix} \hat{a}'_1 \\ \hat{a}'_2 \end{pmatrix} \hat{B}^\dagger = \begin{pmatrix} B_{11}^* & B_{21}^* \\ B_{12}^* & B_{22}^* \end{pmatrix} \begin{pmatrix} \hat{a}'_1 \\ \hat{a}'_2 \end{pmatrix}. \tag{1.190}$$

We can get rid of prime ($'$) in above equation and get

$$\hat{B} \begin{pmatrix} \hat{a}_1 \\ \hat{a}_2 \end{pmatrix} \hat{B}^\dagger = \begin{pmatrix} B_{11}^* & B_{21}^* \\ B_{12}^* & B_{22}^* \end{pmatrix} \begin{pmatrix} \hat{a}_1 \\ \hat{a}_2 \end{pmatrix}. \tag{1.191}$$

We take dagger ($\dagger$) of this equation and get

$$\hat{B} \begin{pmatrix} \hat{a}_1^\dagger \\ \hat{a}_2^\dagger \end{pmatrix} \hat{B}^\dagger = \begin{pmatrix} B_{11} & B_{21} \\ B_{12} & B_{22} \end{pmatrix} \begin{pmatrix} \hat{a}_1^\dagger \\ \hat{a}_2^\dagger \end{pmatrix}. \tag{1.192}$$

By using the result above we can thus check the second equality of Eq. (1.185):

$$\begin{aligned} \hat{B}|n_1, n_2\rangle &= \frac{1}{\sqrt{n_1! n_2!}} (B_{11}\hat{a}_1^\dagger + B_{21}\hat{a}_2^\dagger)^{n_1} (B_{12}\hat{a}_1^\dagger + B_{22}\hat{a}_2^\dagger)^{n_2} |0, 0\rangle \\ &= \frac{1}{\sqrt{n_1! n_2!}} \sum_{k_1=0}^{n_1} \sum_{k_2=0}^{n_2} \binom{n_1}{k_1} \binom{n_2}{k_2} (B_{11})^{k_1} (B_{21})^{n_1-k_1} \\ &\quad \times (B_{12})^{k_2} (B_{22})^{n_2-k_2} \sqrt{(k_1+k_2)!(n_1+n_2-k_1-k_2)!} \\ &\quad \times |k_1+k_2, n_1+n_2-k_1-k_2\rangle. \end{aligned} \tag{1.193}$$

Fig. 1.54 A coherent state
$|\alpha\rangle$ and a vacuum $|0\rangle$ enter to
a 50/50 beam splitter

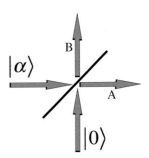

It is obvious from this equation that in any configuration of n_1 and n_2 when the photon number of one of the output beams is $(k_1 + k_2)$ then the other one is $(n_1 + n_2 - k_1 - k_2)$, which corresponds to energy conservation, and the overall state is a superposition of those. In short, in any configuration of n_1 and n_2 the two output beams are entangled.

Since photon-number states $|n\rangle$ are basis states, any state can be described by a superposition of them. So one might think that we could create entanglement with any states of inputs. That is however not true. For example, we consider the case of Fig. 1.54, where a coherent state $|\alpha\rangle$ and a vacuum $|0\rangle$ enter a 50/50 beam splitter. The matrix of a 50/50 beam splitter is

$$\underline{B}_{50/50} = \frac{1}{\sqrt{2}} \begin{pmatrix} 1 & -1 \\ 1 & 1 \end{pmatrix}. \tag{1.194}$$

By using this equation and Eq. (1.193) we calculate the output from a 50/50 beam splitter when a photon-number state n and a vacuum enter the beam splitter.

$$
\begin{aligned}
\hat{B}|n, 0\rangle &= \frac{1}{\sqrt{n!}} \sum_{k=0}^{n} \binom{n}{k} \left(\frac{1}{\sqrt{2}}\right)^k \left(\frac{1}{\sqrt{2}}\right)^{n-k} \sqrt{k!(n-k)!}|k, n-k\rangle \\
&= \sum_{k=0}^{n} \sqrt{\binom{n}{k}} \left(\frac{1}{\sqrt{2}}\right)^n |k, n-k\rangle.
\end{aligned}
\tag{1.195}
$$

By using this equation we can calculate the output state when a coherent state $|\alpha\rangle$ and a vacuum $|0\rangle$ enter the 50/50 beam splitter as follows:

$$
\begin{aligned}
\hat{B}_{50/50}|\alpha, 0\rangle &= \hat{B}_{50/50} e^{-\frac{|\alpha|^2}{2}} \sum_{n=0}^{\infty} \frac{\alpha^n}{\sqrt{n!}} |n, 0\rangle \\
&= e^{-\frac{|\alpha|^2}{2}} \sum_{n=0}^{\infty} \frac{\alpha^n}{\sqrt{n!}} \sum_{k=0}^{n} \sqrt{\binom{n}{k}} \left(\frac{1}{\sqrt{2}}\right)^n |k, n-k\rangle \\
&= e^{-\frac{|\alpha|^2}{2}} \sum_{n=0}^{\infty} \sum_{k=0}^{n} \sqrt{\frac{1}{k!(n-k)!}} \left(\frac{\alpha}{\sqrt{2}}\right)^n |k, n-k\rangle
\end{aligned}
$$

$$= e^{-\frac{|\alpha|^2}{2}} \sum_{k=0}^{\infty} \sum_{m=0}^{\infty} \frac{\left(\frac{\alpha}{\sqrt{2}}\right)^k}{\sqrt{k!}} |k\rangle_A \otimes \frac{\left(\frac{\alpha}{\sqrt{2}}\right)^m}{\sqrt{m!}} |m\rangle_B$$

$$= e^{-\frac{\left|\frac{\alpha}{\sqrt{2}}\right|^2}{2}} \sum_{k=0}^{\infty} \frac{\left(\frac{\alpha}{\sqrt{2}}\right)^k}{\sqrt{k!}} |k\rangle_A \otimes e^{-\frac{\left|\frac{\alpha}{\sqrt{2}}\right|^2}{2}} \sum_{m=0}^{\infty} \frac{\left(\frac{\alpha}{\sqrt{2}}\right)^m}{\sqrt{m!}} |m\rangle_B$$

$$= |\alpha/\sqrt{2}\rangle_A \otimes |\alpha/\sqrt{2}\rangle_B. \tag{1.196}$$

From the result we can see that the two output beams are not entangled.

The reason for this is that a coherent state is a classical state even though it is superposition of photon-number state. On the other hand, photon-number states are nonclassical states, and that is why these states can create entanglement. We need non-classical resources to create entanglement.

From now on we will explain two historical examples of quantum entanglement. The two examples are the Hong–Ou–Mandel effect and the Einstein–Podolsky–Rosen (EPR) state. The Hong–Ou–Mandel effect is the first example of a quantum-mechanical effect of photons and the EPR state originates from the paradox proposed by Einstein et al. to elucidate the incompleteness of quantum mechanics.

First we explain the Hong–Ou–Mandel effect. Figure 1.55 shows the situation for the Hong–Ou–Mandel effect, where two single photon states $|1\rangle$ enter a 50/50 beam splitter from both sides. We can calculate the state $\hat{B}_{50:50}|1, 1\rangle$ by using Eq. (1.193) as follows:

$$\hat{B}_{50:50}|1, 1\rangle = \sum_{k_1=0}^{1} \sum_{k_2=0}^{1} \left(\frac{1}{\sqrt{2}}\right)^{k_1} \left(\frac{1}{\sqrt{2}}\right)^{1-k_1} \left(-\frac{1}{\sqrt{2}}\right)^{k_2} \left(\frac{1}{\sqrt{2}}\right)^{1-k_2}$$

$$\times \sqrt{(k_1 + k_2)!(2 - k_1 - k_2)!} \, |k_1 + k_2, 2 - k_1 - k_2\rangle,$$

$$= \frac{1}{\sqrt{2}} \cdot \frac{1}{\sqrt{2}} \sqrt{2!} \, |0, 2\rangle$$

$$+ \frac{1}{\sqrt{2}} \cdot \left(-\frac{1}{\sqrt{2}}\right) \sqrt{1!1!} \, |1, 1\rangle$$

Fig. 1.55 Two single-photon states $|1\rangle$ enter a 50/50 beam splitter from both sides

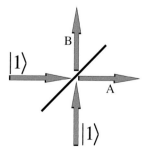

$$+ \frac{1}{\sqrt{2}} \cdot \frac{1}{\sqrt{2}} \sqrt{1!1!} \, |1, 1\rangle$$

$$+ \frac{1}{\sqrt{2}} \cdot \left(-\frac{1}{\sqrt{2}}\right) \sqrt{2!} \, |2, 0\rangle,$$

$$= \frac{1}{\sqrt{2}} \big(|0, 2\rangle - |2, 0\rangle\big). \tag{1.197}$$

Of course, the output beams A and B are entangled, but the most striking fact is that there is no probability of $|1, 1\rangle$, which is caused by quantum interference. It means that there is no probability that two single photons simultaneously exist in both beams A and B. It cannot happen in classical mechanics but can happen in quantum mechanics. This effect is called the Hong–Ou–Mandel effect, named after the people who verified it. It elucidates that photons are not classical particles but quanta and is sometimes called "bunching" of photons. It is because photons stick together instead of being separated when they meet at a beam splitter.

Another important example of entanglement is the EPR state. Einstein, Podolsky, and Rosen proposed the following EPR state to elucidate the incompleteness of quantum mechanics. Position operator $\hat{x}$ and momentum operator $\hat{p}$ for quantum A do not commute and satisfy $[\hat{x}_A, \hat{p}_A] = i\hbar$. Similarly $[\hat{x}_B, \hat{p}_B] = i\hbar$ holds for quantum B. Naively speaking there is the uncertainty relationship between the position and momentum for quanta A and B. On the other hand, "global" operators $\hat{x}_A - \hat{x}_B$ and $\hat{p}_A + \hat{p}_B$ do commute as follows:

$$[\hat{x}_A - \hat{x}_B, \hat{p}_A + \hat{p}_B] = [\hat{x}_A, \hat{p}_A] + [\hat{x}_A, \hat{p}_B] - [\hat{x}_B, \hat{p}_A] - [\hat{x}_B, \hat{p}_B]$$
$$= i\hbar - i\hbar$$
$$= 0. \tag{1.198}$$

So there should be simultaneous eigenstates of these global operators. When both of the eigenvalues are zero ($\hat{x}_A - \hat{x}_B \to 0, \hat{p}_A + \hat{p}_B \to 0$), the eigenstate is called the EPR state $|\text{EPR}\rangle$. The definition is the following:

$$(\hat{x}_A - \hat{x}_B)|\text{EPR}\rangle = 0, \tag{1.199}$$

$$(\hat{p}_A + \hat{p}_B)|\text{EPR}\rangle = 0. \tag{1.200}$$

When we describe the EPR state $|\text{EPR}\rangle$ with eigenstates of $\hat{x}$ ($|x\rangle$), we get

$$|\text{EPR}\rangle = \int_{-\infty}^{\infty} dx |x\rangle_A |x\rangle_B. \tag{1.201}$$

We can easily understand the equation because $\hat{x}_A - \hat{x}_B \to 0$. Furthermore, we can see that quanta A and B are entangled because the equation cannot be described with a simple tensor product of state vectors of quanta A and B.

We can change the basis from $|x\rangle$ to $|p\rangle$ and get

$$\int_{-\infty}^{\infty} dx|x\rangle_A |x\rangle_B$$

$$= \int_{-\infty}^{\infty} dx \int_{-\infty}^{\infty} dp_A |p_A\rangle_A \,_A\langle p_A|x\rangle_A \otimes \int_{-\infty}^{\infty} dp_B |p_B\rangle_B \,_B\langle p_B|x\rangle_B$$

$$= \int_{-\infty}^{\infty} dx \int_{-\infty}^{\infty} dp_A \frac{1}{\sqrt{2\pi\hbar}} e^{-\frac{i}{\hbar} x p_A} |p_A\rangle_A \otimes \int_{-\infty}^{\infty} dp_B \frac{1}{\sqrt{2\pi\hbar}} e^{-\frac{i}{\hbar} x p_B} |p_B\rangle_B$$

$$= \int_{-\infty}^{\infty} \int_{-\infty}^{\infty} dp_A dp_B \frac{1}{2\pi\hbar} \int_{-\infty}^{\infty} dx \, e^{-\frac{i}{\hbar} x (p_A + p_B)} |p_A\rangle_A |p_B\rangle_B,$$

$$= \int_{-\infty}^{\infty} \int_{-\infty}^{\infty} dp_A dp_B \delta(p_A + p_B) |p_A\rangle_A |p_B\rangle_B$$

$$= \int_{-\infty}^{\infty} dp_A |p_A\rangle_A | - p_A\rangle_B$$

$$= \int_{-\infty}^{\infty} dp|p\rangle_A | - p\rangle_B. \tag{1.202}$$

Here we used the equations

$$\delta(\hat{p}_A - \hat{p}_B) = \frac{1}{2\pi} \int_{-\infty}^{\infty} dx \, e^{\pm ix(p_A - p_B)}, \tag{1.203}$$

and

$$\hat{I} = \int_{-\infty}^{\infty} dp|p\rangle\langle p|. \tag{1.204}$$

In Eq. (1.202) we can see that the EPR state is also an eigenstate of $\hat{p}_A + \hat{p}_B$ with an eigenvalue of zero. Of course, since quanta A and B are entangled, the state cannot be described with a simple tensor product of state vectors of quanta A and B even after changing the basis from $|x\rangle$ to $|p\rangle$.

On top of that, we can show that the entanglement is intact after changing from a continuous basis to a discrete basis with

$$\int_{-\infty}^{\infty} dx|x\rangle_A |x\rangle_B = \sum_{n=0}^{\infty} |n\rangle_A |n\rangle_B. \tag{1.205}$$

We can check this in the following way:

$$\int_{-\infty}^{\infty} dx|x\rangle_A |x\rangle_B$$

$$= \int_{-\infty}^{\infty} dx \sum_{n_A=0}^{\infty} |n\rangle_A \,_A\langle n_A|x\rangle_A \otimes \sum_{n_B=0}^{\infty} |n\rangle_B \,_B\langle n_B|x\rangle_B$$

$$= \int_{-\infty}^{\infty} dx \sum_{n_A=0}^{\infty} \sqrt{\frac{1}{\sqrt{\pi}2^{n_A}n_A!}} H_{n_A}(x)e^{-\frac{x^2}{2}} |n_A\rangle_A$$

$$\otimes \sum_{n_B=0}^{\infty} \sqrt{\frac{1}{\sqrt{\pi}2^{n_B}n_B!}} H_{n_B}(x)e^{-\frac{x^2}{2}} |n_B\rangle_B$$

$$= \frac{1}{\sqrt{\pi}} \sum_{n_A=0}^{\infty} \sum_{n_B=0}^{\infty} \sqrt{\frac{1}{2^{n_A+n_B}n_A!n_B!}} \int_{-\infty}^{\infty} dx H_{n_A}(x)H_{n_B}(x)e^{-x^2} |n_A\rangle_A |n_B\rangle_B$$

$$= \sum_{n=0}^{\infty} |n\rangle_A |n\rangle_B. \tag{1.206}$$

Here $H_n(x)$ are the Hermite polynomials, $\langle n|x\rangle = \varphi_n(x)$, and we set $m = 1$, $\omega = 1$, and $\hbar = 1^2$ in Eq. (1.98). We also used the formula

$$\int_{-\infty}^{\infty} dx H_{n_A}(x)H_{n_B}(x)e^{-x^2} = \sqrt{\pi}\delta_{n_A,n_B}2^{n_A}n_A!. \tag{1.207}$$

Thus, we can describe the entanglement of the EPR state $|\text{EPR}\rangle$ with continuous and discrete bases. There is one caveat. The EPR state $|\text{EPR}\rangle$ is an unphysical state, because it cannot be normalized. So we have to think about a physical state which asymptotically approaches the EPR state $|\text{EPR}\rangle$ in an extreme condition. That is a two-mode squeezed vacuum $|\text{EPR}*\rangle$.

A two-mode squeezed vacuum $|\text{EPR}*\rangle$ is created by two independent squeezed vacua and a 50/50 beam splitter as shown in Fig. 1.56. Here we take a beam splitter matrix $\underline{B}_{50/50*}$ for that purpose:

$$\underline{B}_{50/50*} = \frac{1}{\sqrt{2}} \begin{pmatrix} 1 & 1 \\ 1 & -1 \end{pmatrix}. \tag{1.208}$$

So the Heisenberg picture of the beam splitter transformation should be

Fig. 1.56 Squeezed vacua $\hat{S}(r)|0\rangle$ and $\hat{S}(-r)|0\rangle$ enter a 50/50 beam splitter. The output state is a two-mode squeezed vacuum $|\text{EPR}*\rangle$

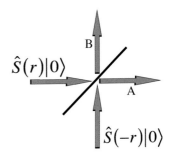

$\hat{S}(r)|0\rangle$

B

A

$\hat{S}(-r)|0\rangle$

---

[2]We can use $\hbar = 1/2$ instead of $\hbar = 1$.

$$\hat{B}_{50/50*}^{\dagger} \begin{pmatrix} \hat{a}_A \\ \hat{a}_B \end{pmatrix} \hat{B}_{50/50*} = \hat{B}_{50/50*} \begin{pmatrix} \hat{a}_A \\ \hat{a}_B \end{pmatrix} \hat{B}_{50/50*}^{\dagger}$$

$$= \frac{1}{\sqrt{2}} \begin{pmatrix} 1 & 1 \\ 1 & -1 \end{pmatrix} \begin{pmatrix} \hat{a}_A \\ \hat{a}_B \end{pmatrix}. \qquad (1.209)$$

The reason why we use the beam splitter matrix above is that the inverse transformation, i.e., the inverse matrix, is the same as the original one and everything becomes simpler.

Let's calculate the output state $|EPR*\rangle_{AB}$:

$$\begin{aligned} |EPR*\rangle_{AB} &= \hat{B}_{50/50*} \hat{S}_A(r) |0\rangle_A \otimes \hat{S}_B(-r) |0\rangle_B \\ &= \hat{B}_{50/50*} \exp\left[\frac{r}{2}(\hat{a}_A^{\dagger 2} - \hat{a}_A^2)\right] \exp\left[\frac{r}{2}(\hat{a}_B^2 - \hat{a}_B^{\dagger 2})\right] |0\rangle_A \otimes |0\rangle_B \\ &= \hat{B}_{50/50*} \exp\left[\frac{r}{2}(\hat{a}_A^{\dagger 2} - \hat{a}_A^2 + \hat{a}_B^2 - \hat{a}_B^{\dagger 2})\right] \hat{B}_{50/50*}^{\dagger} |0\rangle_A \otimes |0\rangle_B \\ &= \exp\left[r(\hat{a}_A^{\dagger} \hat{a}_B^{\dagger} - \hat{a}_A \hat{a}_B)\right] |0\rangle_A \otimes |0\rangle_B \\ &= \sqrt{1-q^2} \sum_{n=0}^{\infty} q^n |n\rangle_A \otimes |n\rangle_B. \qquad (1.210) \end{aligned}$$

Here $q = \tanh r$ and we used the following equation for the last equality.

$$\begin{aligned} &\exp\left[r(\hat{a}_A^{\dagger} \hat{a}_B^{\dagger} - \hat{a}_A \hat{a}_B)\right] \\ &= \exp\left(\hat{a}_A^{\dagger} \hat{a}_B^{\dagger} \tanh r\right) \exp\left[-(\hat{a}_A^{\dagger} \hat{a}_A + \hat{a}_B^{\dagger} \hat{a}_B + 1) \ln(\cosh r)\right] \\ &\quad \times \exp\left(-\hat{a}_A \hat{a}_B \tanh r\right). \qquad (1.211) \end{aligned}$$

It is obvious from Eq. (1.210) that the two output beams are entangled. Although the exception is the case of $r = 0$, where the output state becomes $|0\rangle_A \otimes |0\rangle_B$. It is trivial because this case corresponds to both of input beams are vacua.

Here the transformation $\exp\left[r(\hat{a}_A^{\dagger} \hat{a}_B^{\dagger} - \hat{a}_A \hat{a}_B)\right]$ in Eq. (1.210) is called "two-mode squeezing". So $|EPR*\rangle_{AB}$ is called a "two-mode squeezed vacuum". In the limit of $r \to \infty$ ($q \to 1$) a two-mode squeezed vacuum becomes the EPR state $|EPR\rangle$. Of course, it is obvious that we cannot create a perfect EPR state because we need infinite amount of energy for $r \to \infty$.

Chapter 2
Creation of Quantum States of Light

In the previous chapter we described various aspects of quantum states of light. Following a historical order we will now explain how to create quantum states of light, including coherent states, squeezed states, a Schrödinger's cat state, a single-photon state, and a superposition of photon-number states. In the end of this chapter we will explain how to create entanglement.

2.1 Creation of Coherent States of Light

Coherent states of light can in principle be created by a laser. Laser oscillation occurs through amplification of spontaneous emission from the laser material, which corresponds to a vacuum $|0\rangle$. We will not explain in detail, but the process corresponds to the displacement operation $\hat{D}(\alpha)$ in phase space, as shown in Eqs. (1.127) and (1.129). In Eq. (1.129) we used

$$\hat{D}(\alpha) = e^{-ix_0 p_0} e^{i2p_0 \hat{x}} e^{-i2x_0 \hat{p}}, \tag{2.1}$$

which is equivalent with

$$\hat{D}(\alpha) = e^{\alpha \hat{a}^\dagger - \alpha^* \hat{a}}. \tag{2.2}$$

Let's make the displacement operation for a vacuum $|0\rangle$:

$$\begin{aligned}
\hat{D}(\alpha)|0\rangle &= e^{\alpha \hat{a}^\dagger - \alpha^* \hat{a}}|0\rangle \\
&= e^{\alpha \hat{a}^\dagger} e^{-\alpha^* \hat{a}} e^{-\frac{1}{2}[\alpha \hat{a}^\dagger, -\alpha^* \hat{a}]}|0\rangle \\
&= e^{-\frac{|\alpha|^2}{2}} e^{\alpha \hat{a}^\dagger} e^{-\alpha^* \hat{a}}|0\rangle \\
&= e^{-\frac{|\alpha|^2}{2}} e^{\alpha \hat{a}^\dagger}|0\rangle
\end{aligned}$$

© The Author(s) 2015
A. Furusawa, *Quantum States of Light*, SpringerBriefs
in Mathematical Physics, DOI 10.1007/978-4-431-55960-3_2

Fig. 2.1 A coherent state
$|\alpha\rangle$ is created from a vacuum
$|0\rangle$ through displacement
operation $\hat{D}(\alpha)$

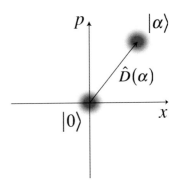

$$= e^{-\frac{|\alpha|^2}{2}} \sum_{n=0}^{\infty} \frac{\alpha^n}{n!} \hat{a}^{\dagger n} |0\rangle$$

$$= e^{-\frac{|\alpha|^2}{2}} \sum_{n=0}^{\infty} \frac{\alpha^n}{n!} \sqrt{n!} |n\rangle$$

$$= e^{-\frac{|\alpha|^2}{2}} \sum_{n=0}^{\infty} \frac{\alpha^n}{\sqrt{n!}} |n\rangle$$

$$= |\alpha\rangle. \tag{2.3}$$

Here we used the formula

$$e^{\hat{A}+\hat{B}} = e^{\hat{A}} e^{\hat{B}} e^{-[\hat{A},\hat{B}]/2}. \tag{2.4}$$

Note that this formula holds only when $[\hat{A}, [\hat{A}, \hat{B}]] = [\hat{B}, [\hat{A}, \hat{B}]] = 0$. Figure 2.1 illustrates the displacement operation.

If we take a Hamiltonian for the displacement operation as

$$\hat{H}_\alpha \propto i(\alpha \hat{a}^\dagger - \alpha^* \hat{a}), \tag{2.5}$$

then the "time evolution" of the quantum complex amplitude or the annihilation operator $\hat{a}$ can be described as

$$e^{i\frac{\hat{H}_\alpha}{\hbar}t} \hat{a} e^{-i\frac{\hat{H}_\alpha}{\hbar}t}. \tag{2.6}$$

By using this and Eq. (1.160) we can calculate the "time evolution" of $\hat{a}$ as

$$\hat{D}^\dagger(\alpha)\hat{a}\hat{D}(\alpha) = e^{-(\alpha \hat{a}^\dagger - \alpha^* \hat{a})} \hat{a} e^{\alpha \hat{a}^\dagger - \alpha^* \hat{a}}$$

$$= \hat{a} - [\alpha \hat{a}^\dagger - \alpha^* \hat{a}, \hat{a}] + \frac{1}{2!}[\alpha \hat{a}^\dagger - \alpha^* \hat{a}, [\alpha \hat{a}^\dagger - \alpha^* \hat{a}, \hat{a}]] + \cdots$$

$$= \hat{a} + \alpha. \tag{2.7}$$

Fig. 2.2 Phase diffusion of
an actual laser

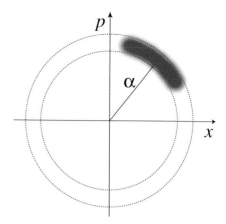

Here we can see that the quantum complex amplitude or the annihilation operator $\hat{a}$ changes to $\hat{a} + \alpha$ through the displacement operation $\hat{D}(\alpha)$. This situation can be understood from Fig. 2.1. In short, the displacement operation $\hat{D}(\alpha)$ corresponds to laser oscillation itself when it acts on a vacuum $|0\rangle$. This situation can be seen in the Hamiltonian of the displacement operation $(\hat{H}_\alpha)$ given in Eq. (2.5). In Eq. (2.5) can $\alpha\hat{a}^\dagger$ and $\alpha^*\hat{a}$ be regarded as stimulated emission and absorption, respectively.

In the case of an actual laser there is phase diffusion on top of stimulated emission and absorption. So the actual picture of a laser should be the one shown in Fig. 2.2. Although there are even some technical (classical) noises on top of the phase diffusion, the quantum noise is dominant in the high-frequency region. Also, when we want to cancel out the phase diffusion, we use a coherent beam as the reference. More precisely, we usually use a coherent beam from the same laser for the local oscillator light for homodyne measurement to cancel out the phase diffusion.

There is another way to create coherent states of light. We can create coherent states of light at the sideband of the laser frequency by using amplitude or phase modulators. The advantage of this method is that technical (classical) noises are intrinsically very small at higher frequencies and it is easy to realize a situation in which the quantum noise is dominant. So we can create an ideal coherent state with this method. As an example we show an experimental result of quadrature amplitude measurement of a coherent state which is created with this method. Laser light was divided into two beams, and one of the beams was used for a local oscillator beam and the other was modulated to create a coherent state at the modulation sideband. Figure 2.3 shows the experimental phase dependence of amplitude for the coherent state. We can see that it is qualitatively a coherent state. As a reference we show the result of the same experiment for a vacuum $|0\rangle$ in Fig. 2.4. We can see that the quantum fluctuations in Figs. 2.3 and 2.4 look similar. More quantitatively we can calculate the photon-number distribution for the coherent state with the data presented in Fig. 2.3. The result is shown in Fig. 2.5. We can see a Poisson distribution there, which is supposed to follow Eq. (1.40). Finally we show a Wigner function calculated from Fig. 2.3 as Fig. 2.6.

Fig. 2.3 Experimental result
of phase dependence of
amplitude for a coherent
state

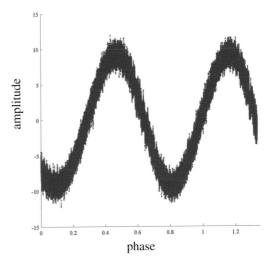

Fig. 2.4 Experimental result
of phase dependence of
amplitude for a vacuum

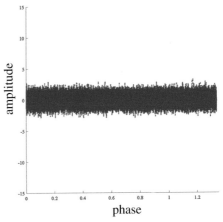

Fig. 2.5 Photon-number
distribution calculated with
the data presented in Fig. 2.3

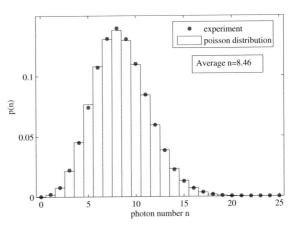

Fig. 2.6 Wigner function calculated with the data presented in Fig. 2.3

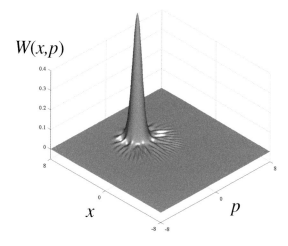

$W(x,p)$

Fig. 2.7 An intensity-fluctuating beam enters a 50/50 beam splitter

Intensity change

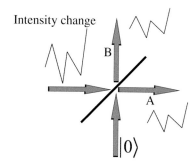

    We consider quantum fluctuation here. We showed an experimental result of phase dependence of amplitude for a vacuum $|0\rangle$ in Fig. 2.4, which corresponds to quantum fluctuation or quantum noise. The experiment was done with balanced homodyne measurement, where measured light mode $\hat{a}_1$ was in a vacuum state $|0\rangle$, i.e., no input and the other input $\hat{a}_2$ was used for a local oscillator in a coherent state $|\alpha\rangle$. It might seem funny at a first glance. The reason is the following. Some light beam enters a 50/50 beam splitter. It can be in a coherent state but it can be in other states as well. In any case, the two output beams have the same intensity fluctuations. So naively speaking the output of the balanced homodyne measurement should be zero, because the same signals should be totally cancelled out when we took the difference signal of two homodyne currents. However, we got Fig. 2.4. What happened here? Of course, it can be easily solved when we use the particle picture of photons (Fig. 2.7).
    We can explain the situation with Fig. 2.8b. It is the particle picture. In this picture individual photons have no correlation and randomly enter the 50/50 beam splitter. Each photon has a 50 %-chance of transmission and a 50 %-chance of reflection at the beam splitter. Of course, a reflected photon cannot be transmitted and a transmitted photon cannot be reflected. So the output two beams have opposite-phase intensity noises. Since we take the difference of them to get a balanced homodyne signal, we

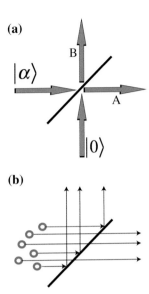

(a)

(b)

consequently take the sum of the opposite-phase intensity noises. As a result, we can
get the noise originated from the particle picture of photons, which corresponds to
the quantum noise or vacuum noise shown in Fig. 2.4. In other words, the existence
of this noise is a proof of the existence of photons. Moreover, since the photon's
(particle's) temporal mode is a time-domain delta function $\delta(t)$, the noise spectrum
becomes uniform or white in frequency domain. Thus, it is often called "shot noise".

2.2 Creation of a Squeezed Vacuum

A coherent state can be regarded as a classical electro-magnetic field plus quantum
noise, which corresponds to a classical state of light in some sense. So the "true"
quantum state of light created for the very first time would be a squeezed vacuum
$\hat{S}(r)|0\rangle$. As mentioned in the preface, although it was created by Slusher et al. in
1985 [1] for the very first time with a third-order nonlinear process, most people
nowadays use a degenerate parametric process (second-order nonlinearity), which
was realized by Wu et al. in 1986 [2]. It is because the parametric process is far more
efficient than the third-order nonlinear process. We will thus explain the degenerate
parametric process in this section.

Figure 2.9 shows the schematic of the degenerate parametric process. In the
process, strong pump light of frequency 2ω (complex amplitude E_3) and input light of
frequency ω (complex amplitude E_{in}) are coupled in a second-order nonlinear crystal ($\chi^{(2)}$) and create the difference frequency component of frequency $\omega = 2\omega - \omega$
(complex amplitude E_2). This component couples coherently with the input light
and creates the E_{out} output. The input-output relation is

Fig. 2.9 Degenerate
parametric process

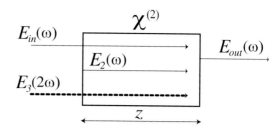

Fig. 2.10 Degenerate
optical parametric process
with a cavity, which
corresponds to the
Bogoliubov transformation,
i.e., squeezing operation

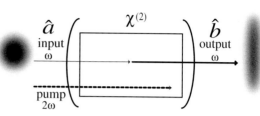

$$E_{\text{out}} = E_{\text{in}} \cosh r + E_{\text{in}}^* \sinh r, \tag{2.8}$$

where

$$r = \frac{z}{2c} \cdot \frac{\omega}{n} \left| \chi^{(2)} E_3 \right|, \tag{2.9}$$

r is the squeezing parameter, z is the length of the nonlinear crystal, c is the speed
of light, n is the refractive index of the nonlinear crystal, and $\chi^{(2)}$ is the nonlinear
coefficient.

From $|\cosh r|^2 - |\sinh r|^2 = 1$, we can see that Eq. (2.8) corresponds to the
Bogoliubov transformation in Eqs. (1.154) or (1.162). When we take $\hat{a}$ as a complex
amplitude of a light field, Eq. (2.8) is the Bogoliubov transformation itself. We can
make a Bogoliubov transformation by using a degenerate parametric process. Note
that we cannot get very big r when we use ordinary nonlinear crystals. So we usually use a cavity to enhance the nonlinearity to get bigger r, which corresponds to
lengthening the effective crystal length z. A schematic of the process is shown in
Fig. 2.10.

We usually create a squeezed vacuum $\hat{S}(r)|0\rangle$ in actual experiments. It is because
we can have a vacuum input $|0\rangle$ even without preparing some input state. Although
we inevitably will suffer from losses, e.g., coupling losses in the case of some actual
input state, we do not have to be worried about losses in the case of a vacuum input.
Losses always introduce a vacuum, that always makes the coupling efficiency 100 %
for a vacuum input! So it is easy to make a high-level squeezing operation when the
input is a vacuum. Moreover, squeezing parameter r should go to infinity depending
on the pump power as can be seen in Eq. (2.9). However, the observable r is not so
big in the real experiments. We will explain the reason below.

We have to make a balanced homodyne measurement to detect squeezed light, which was explained in Sect. 1.3. It is because a balanced homodyne measurement is a phase-sensitive measurement and we can independently measure squeezed and anti-squeezed components. We use the setup shown in Fig. 1.4 for homodyne measurements and obtain $\langle \hat{I}_2 - \hat{I}_1 \rangle$ of Eq. (1.69) to get the marginal distributions. To get the squeezing parameter r we usually obtain $\langle (\hat{I}_2 - \hat{I}_1)^2 \rangle$ by using a spectrum analyzer. In that case, we use Eq. (1.69) and get

$$\langle (\hat{I}_2 - \hat{I}_1)^2 \rangle = 4|\alpha|^2 \cdot {}_1\langle \psi | (\hat{x}_1 \cos \theta + \hat{p}_1 \sin \theta)^2 | \psi \rangle_1. \tag{2.10}$$

Since the input state is a squeezed vacuum $\hat{S}(r)|0\rangle$, then $|\psi\rangle_1 = \hat{S}(r)|0\rangle_1$. So the output of the spectrum analyzer should be

$$\begin{aligned}
\langle (\hat{I}_2 - \hat{I}_1)^2 \rangle &= 4|\alpha|^2 \cdot {}_1\langle 0|\hat{S}_1^\dagger(r)(\hat{x}_1 \cos \theta + \hat{p}_1 \sin \theta)^2 \hat{S}_1(r)|0\rangle_1 \\
&= 4|\alpha|^2 \Big[{}_1\langle 0|\hat{S}_1^\dagger(r)\hat{x}_1^2 \hat{S}_1(r)|0\rangle_1 \cos^2 \theta \\
&\quad + {}_1\langle 0|\hat{S}_1^\dagger(r)\hat{p}_1^2 \hat{S}_1(r)|0\rangle_1 \sin^2 \theta \\
&\quad + {}_1\langle 0|\hat{S}_1^\dagger(r)(\hat{x}_1 \hat{p}_1 + \hat{p}_1 \hat{x}_1)\hat{S}_1(r)|0\rangle_1 \sin \theta \cos \theta \Big] \\
&= 4|\alpha|^2 \Big({}_1\langle 0|\hat{x}_1^2 e^{-2r}|0\rangle_1 \cos^2 \theta + {}_1\langle 0|\hat{p}_1^2 e^{2r}|0\rangle_1 \sin^2 \theta \Big) \\
&= |\alpha|^2 (e^{-2r} \cos^2 \theta + e^{2r} \sin^2 \theta). \tag{2.11}
\end{aligned}$$

Here we used Eqs. (1.165)–(1.168). From the result we should see a periodic structure corresponding to squeezed and anti-squeezed quadratures (e^{-2r} and e^{2r}) depending on the phase of the local-oscillator beam for the balanced homodyne measurement. Here e^{-2r} and e^{2r} correspond to variances $\langle (\Delta x_1)^2 \rangle$ and $\langle (\Delta p_1)^2 \rangle$. It is because ${}_1\langle 0|\hat{S}_1^\dagger(r)\hat{x}_1 \hat{S}_1(r)|0\rangle_1 = {}_1\langle 0|\hat{S}_1^\dagger(r)\hat{p}_1 \hat{S}_1(r)|0\rangle_1 = 0$ which corresponds to that the averaged amplitude is zero.

As shown above, we can measure the squeezing level and determine the squeezing parameter r. Now we have to think about the influence losses have on squeezing. A squeezed vacuum is a superposition of even-photon states as shown in previous sections. When squeezed vacuum is exposed to loss, the even-photon nature or nonclassicality is degraded. Especially the e^{-2r} component is degraded. We will consider the consequences of this below.

We make a model shown in Fig. 2.11 to think about the influence of loss. In this figure we have a fictitious beam splitter (BS) on top of the setup of Fig. 1.4, which represents losses. Here the input for the homodyne measurement is $\hat{a}_1$ and loss is represented by a vacuum $|0\rangle_3$ from the BS. The input after having losses is $\hat{a}_1'$.

Fig. 2.11 Power measurement with a spectrum analyzer (*SA*) for balanced homodyne current. The input beam $\hat{a}_1$ experiences losses, which is represented by mode $\hat{a}_3$ (a vacuum $|0\rangle$) and a fictitious beam splitter (*BS*)

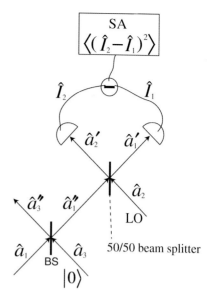

When we set the local-oscillator phase $\theta = 0$, the output from a spectrum analyzer $\langle(\hat{I}_2 - \hat{I}_1)^2\rangle$ can be calculated with Eq. (2.10) as follows:

$$
\begin{aligned}
\langle(\hat{I}_2 - \hat{I}_1)^2\rangle &= 4|\alpha|^2 {}_3\langle 0| \otimes {}_1\langle\psi|e^{-i\Theta\hat{L}_2}\hat{x}_1^2 e^{i\Theta\hat{L}_2}|\psi\rangle_1 \otimes |0\rangle_3 \\
&= 4|\alpha|^2 {}_3\langle 0| \otimes {}_1\langle\psi|[\hat{x}_1^2 \cos^2(\Theta/2) + \hat{x}_3^2 \sin^2(\Theta/2)]|\psi\rangle_1 \otimes |0\rangle_3 \\
&= 4|\alpha|^2 \left[{}_1\langle\psi|\hat{x}_1^2|\psi\rangle_1 \cos^2(\Theta/2) + {}_3\langle 0|\hat{x}_3^2|0\rangle_3 \sin^2(\Theta/2) \right] \\
&= 4|\alpha|^2 \left[{}_1\langle\psi|\hat{x}_1^2|\psi\rangle_1 \cos^2(\Theta/2) + \frac{1}{4}\sin^2(\Theta/2) \right].
\end{aligned}
\tag{2.12}
$$

Here we used Eq. (1.59)[1] and ${}_3\langle 0|\hat{x}_3|0\rangle_3 = 0$. From the result we can see that some portion of the output of the spectrum analyzer is replaced by a vacuum variance of 1/4 by the ratio of the reflectivity of the fictitious beam splitter $\sin^2(\Theta/2)$. Of course, we can calculate the case of non-zero θ, where the result is similar.

In summary, we can obtain the following result from the measurement of a squeezed vacuum with loss;

$$
|\alpha|^2 \left[(e^{-2r}\cos^2\theta + e^{2r}\sin^2\theta)\cos^2(\Theta/2) + \sin^2(\Theta/2) \right].
\tag{2.13}
$$

When the vacuum fluctuation is set to be one, the squeezing level becomes

$$
e^{-2r}\cos^2(\Theta/2) + \sin^2(\Theta/2).
\tag{2.14}
$$

---

[1]We set $\hat{L}_2 = \frac{1}{2i}\left(\hat{a}_1^\dagger\hat{a}_3 - \hat{a}_1\hat{a}_3^\dagger\right)$.

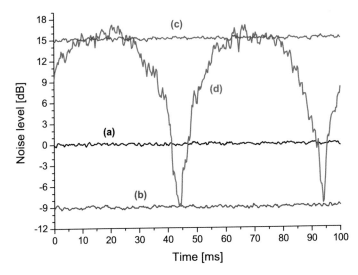

Fig. 2.12 An experimental result of measurements of a squeezed vacuum [3]. It was a world record in squeezing in 2007. **a** Vacuum fluctuation, **b** measurement result when the local oscillator phase was locked at the squeezed quadrature, **c** measurement result when the local oscillator phase was locked at the anti-squeezed quadrature, **d** the local oscillator phase was scanned

The anti-squeezing level becomes

$$e^{2r} \cos^2(\Theta/2) + \sin^2(\Theta/2). \tag{2.15}$$

We will check them with an experimental result below.

Figure 2.12 shows the experimental result from a measurements on a squeezed vacuum. The squeezing level is -9 dB (0.126) and the anti-squeezing level is 15 dB (31.6). It was a world record in squeezing in 2007. We can understand these values by using Eqs. (2.14) and (2.15). In Eq. (2.15) we can neglect the influence of loss. It is because the vacuum fluctuation is minute compared to the anti-squeezing. So we take

$$31.6 \approx e^{2r}, \tag{2.16}$$

and get

$$e^{-2r} = \frac{1}{31.6} \approx 0.0316. \tag{2.17}$$

From Eq. (2.14) we get

$$0.126 \approx 0.0316 \times \cos^2(\Theta/2) + \sin^2(\Theta/2). \tag{2.18}$$

So we get $\sin^2 (\Theta/2) = 0.1$ with $\cos^2 (\Theta/2) + \sin^2 (\Theta/2) = 1$. From these considerations we can conclude that the squeezing level before having a loss was -15 dB (0.0316) and the total loss was 10%. Note that we usually take an observed squeezing level as the experimental squeezing level. So in the above case we usually take the experimental squeezing level as $e^{-2r'} = 0.126$ and the experimental squeezing parameter as $r' = 1.04$.

We mention the history of experimental squeezing level here. The world record in squeezing, reported by Kimble's group at Caltech was -6 dB ($e^{-2r'} = 0.25$) from 1992 to 2006 [4]. Since this record was not broken for 14 years, it became common sense that -6 dB of squeezing would be a physical limit. However, the author's group bravely tried to improve the squeezing level and finally realized -7 dB of squeezing in 2006 [5], and then the world-wide race of chasing a high level of squeezing begun. It is a typical example of "common sense is nonsense".

In 2007 the author's group succeeded in -9 dB of squeezing as shown in Fig. 2.12. In 2008 Schnabel's group at Hannover reported -10 dB [6]. By 2013 a couples of groups in the world reported -13 dB [7]. In any case, the important point was how to reduce pump-induced losses. As previously mentioned, losses degrade the squeezing level. It is trivial now but it was not trivial until 2006.

Before closing this section we show examples of experimental squeezed vacua in various ways. Figure 2.13 shows the phase dependence of amplitude for a squeezed vacuum, Fig. 2.14 shows the photon number distribution calculated with the data presented in Figs. 2.13, and 2.15 shows the Wigner function calculated with the data presented in Fig. 2.13. Figure 2.13 agrees well with the one of theoretical -10 dB of squeezing shown in Fig. 1.51, and we can see the even-photon nature in Fig. 2.14. Figure 2.15 also agrees well with the one for a theoretical Wigner function shown in Fig. 1.50.

Fig. 2.13 An example of phase dependence of amplitude for an experimental squeezed vacuum

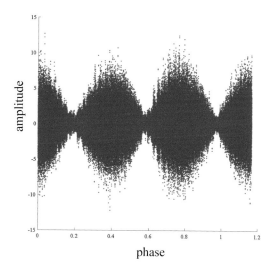

Fig. 2.14 Photon number
distribution calculated with
the data presented in
Fig. 2.13

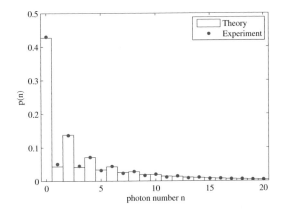

Fig. 2.15 Wigner function
calculated with the data
presented in Fig. 2.13

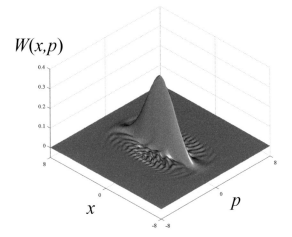

2.3   Creation of a Single-Photon State

There are several ways to generate a single-photon state. However, the only way
to get the negativity of a single-photon Wigner function is the heralded method,
which will be explained below. Of course, in that situation we should see the phase
dependence of amplitude for a single-photon state $|1\rangle$ as shown in Fig. 2.16.

In the case of a heralded way to generate a single-photon state, we use a $-3\,\text{dB}$-
squeezed vacuum $\hat{S}(\ln 2/2)|0\rangle$ in Eq. (1.152). Note that we use a non-degenerate
squeezed vacuum in this case, not an ordinary degenerate one. For the non-degenerate
case, two photons of state $|2\rangle$ are created in different modes A and B, for example two
orthogonally polarized modes. These photon pairs can be created with some particular
phase-matching condition. So Eq. (1.152) is modified and we get the $|\text{PDC}\rangle$ state:

$$|\text{PDC}\rangle = 0.971|0\rangle_A|0\rangle_B + 0.229|1\rangle_A|1\rangle_B + \cdots . \qquad (2.19)$$

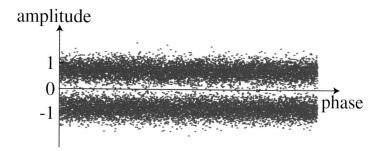

Fig. 2.16 Phase dependence of amplitude for a single-photon state $|1\rangle$ (Fig. 1.12)

This state is a vacuum $|0\rangle$ with a probability of 0.971^2 and is a photon pair A,B with a probability of 0.229^2. That means it is a vacuum $|0\rangle$ most of the time and a photon pair A,B with a very small probability. We sometimes call it "parametric downconversion" (PDC). Of course, it is just a squeezed vacuum with a different phase-matching condition.

A photon pair created by parametric downconversion can be divided into two photons in different paths. For example, we can use a polarization beam splitter when the two photons have orthogonal polarizations. If we detect a photon in beam A for heralding creation of a single-photon state, then we can get a single-photon state $|1\rangle$ in beam B. Figure 2.17 shows the scheme of heralded single-photon creation.

Figure 2.18 shows an example of phase dependence of amplitude for an experimental single-photon state $|1\rangle$. The Wigner function calculated with the data presented in Fig. 2.18 is shown in Fig. 2.19, and the photon-number distribution calculated with the data presented in Fig. 2.18 is shown in Fig. 2.20. We can see the similarity between Figs. 2.16 and 2.18 and reasonable agreement between the theoretical prediction and the experimental result. We can clearly see the negativity of the experimental single-photon Wigner function and it tells us that the state is a highly pure experimental single-photon state. Actually, the portion of calculated single-photon component is more than 80 % as shown in Fig. 2.20.

Fig. 2.17 Scheme of heralded single-photon creation. If we detect a photon in beam *A*, then we can get a single-photon state $|1\rangle$ in beam *B*

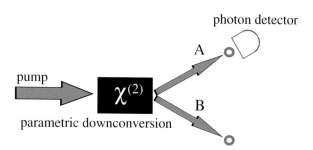

Fig. 2.18 An example of
phase dependence of
amplitude for an
experimental single-photon
state $|1\rangle$. Note that we
usually use $\hbar = 1$ for this
type of plot for some reason

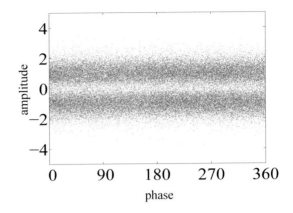

Fig. 2.19 Wigner function
calculated with the data
presented in Fig. 2.18

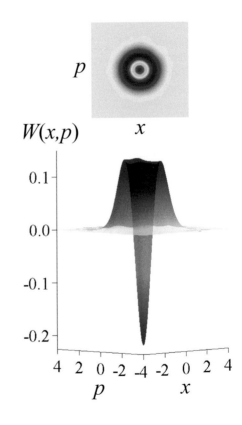

Fig. 2.20 Photon-number distribution calculated with the data presented in Fig. 2.18

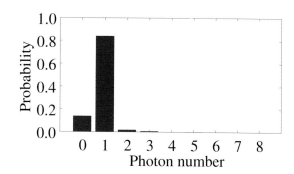

2.4 Creation of a Minus Cat State

We will explain how to create one of the Schrödinger's cat states, a minus cat state, in this section. We consider a heralded creation similar to the previous section. Note that the word "herald" is replaced by "photon subtraction" in this section. As seen in Eq. (1.150), a squeezed vacuum is a superposition of even photons. On the other hand, a minus cat state is a superposition of odd photons. So we can create a minus cat state by subtracting a single photon from a squeezed vacuum.

Let's think about this example. From Eq. (1.151) we can describe a $-4\,$dB-squeezed vacuum $|\text{sqz} - 4\,\text{dB}\rangle$ ($-4\,$dB-squeezing: $e^{-2r} = 0.398$) as

$$|\text{sqz} - 4\,\text{dB}\rangle = 0.950|0\rangle + 0.289|2\rangle + 0.108|4\rangle + 0.0424|6\rangle + \cdots. \quad (2.20)$$

Figure 2.21 shows the phase dependence of amplitude of $-4\,$dB-squeezed vacuum $|\text{sqz} - 4\,\text{dB}\rangle$ calculated with above equation. The envelope of the plot agrees well with the one of a minus cat state in Fig. 1.30.

We subtract a single photon from the $-4\,$dB-squeezed vacuum $|\text{sqz} - 4\,\text{dB}\rangle$. From Eq. (2.20) we get without normalization

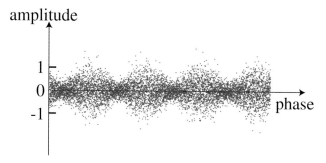

Fig. 2.21 Phase dependence of amplitude of the $-4\,$dB-squeezed vacuum $|\text{sqz} - 4\,\text{dB}\rangle$ calculated with Eq. (2.20)

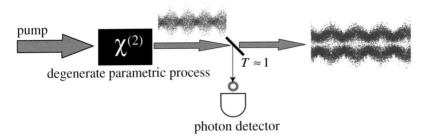

Fig. 2.22 We can create the minus cat state $|\alpha = 1\rangle - |\alpha = -1\rangle$ with single-photon subtraction from the $-4\,\mathrm{dB}$-squeezed vacuum $|\mathrm{sqz} - 4\,\mathrm{dB}\rangle$

$$|\mathrm{sqz} - 4\,\mathrm{dB}(1\,\mathrm{subtract})\rangle = 0.289|1\rangle + 0.108|3\rangle + 0.0424|5\rangle + \cdots . \qquad (2.21)$$

For the normalization we have to divide it by $\sqrt{0.289^2 + 0.108^2 + 0.0424^2 + \cdots} \approx 0.311$. Then we get

$$|\mathrm{sqz} - 4\,\mathrm{dB}(1\,\mathrm{subtract})\rangle \approx 0.929|1\rangle + 0.347|3\rangle + 0.136|5\rangle + \cdots . \qquad (2.22)$$

This state agrees well with the state $|\alpha = 1\rangle - |\alpha = -1\rangle$, whose equation is the following (Eq. (1.119)):

$$N_{\alpha-}(|\alpha = 1\rangle - |\alpha = -1\rangle) = \sqrt{\frac{2e}{e^2 - 1}} \left(|1\rangle + \frac{1}{\sqrt{3!}}|3\rangle + \frac{1}{\sqrt{5!}}|5\rangle + \cdots \right)$$

$$= 0.922|1\rangle + 0.377|3\rangle + 0.00769|5\rangle + \cdots . \qquad (2.23)$$

Actually, when we calculate the phase dependence of amplitude of the state $|\mathrm{sqz} - 4\,\mathrm{dB}(1\,\mathrm{subtract})\rangle$, we cannot see any difference from Fig. 1.30. So we can create the minus cat state $|\alpha = 1\rangle - |\alpha = -1\rangle$ with single-photon subtraction from the $-4\,\mathrm{dB}$-squeezed vacuum $|\mathrm{sqz} - 4\,\mathrm{dB}\rangle$ as shown in Fig. 2.22. The intuitive explanation of the single-photon subtraction is the following. When we detect a photon at the photon detector, we eliminate the probability of no photon or a vacuum $|0\rangle$ for the other output from the $T \sim 1$ beam splitter. So we can eliminate the probability of small amplitude of the $-4\,\mathrm{dB}$-squeezed vacuum $|\mathrm{sqz} - 4\,\mathrm{dB}\rangle$ in Fig. 2.21.

We explain the schematic of Fig. 2.22 a bit in detail. First we create a squeezed vacuum by using a degenerate parametric process. Here we set the squeezing level so as to create a minus cat state as seen above ($-4\,\mathrm{dB}$ or so). Then the squeezed vacuum enters a beam splitter with transmissivity $T \sim 1$. The reflected beam contains one photon at most because the reflectivity is very small. Therefore, when we detect a photon at the photon detector, it follows that we subtract a single photon from the squeezed vacuum. So we can get a minus cat state. The process is similar to heralding creation of a single photon in some sense.

Fig. 2.23 An example of phase dependence of amplitude of an experimental minus cat state $|\alpha = 1\rangle - |\alpha = -1\rangle$, which was created by single-photon subtraction

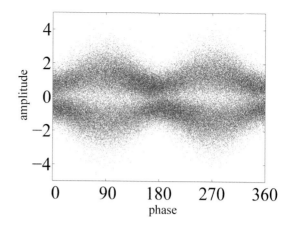

Fig. 2.24 Wigner function calculated with the data presented in Fig. 2.23

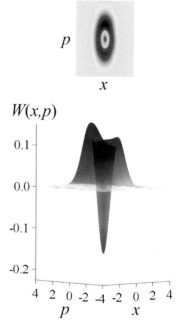

Figure 2.23 shows an example of phase dependence of amplitude of an experimental minus cat state $|\alpha = 1\rangle - |\alpha = -1\rangle$, which was created by single-photon subtraction. The calculated Wigner function with the data presented in Fig. 2.23 is shown in Fig. 2.24 and the calculated photon-number distribution with the data presented in Fig. 2.23 is shown in Fig. 2.25

Figures 2.23 and 1.30 agree well with each other, which means that the experimental result and the theoretical calculation agree well with each other. We can clearly see two opposite-phase coherent states and elimination around zero-amplitude in

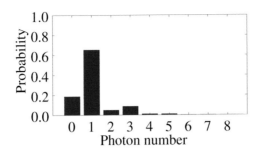

Fig. 2.25 Photon-number distribution calculated with the data presented in Fig. 2.23

the experimental result. Moreover, we can see the negativity and two peaks of the Wigner function in Fig. 2.24, which agree well with Fig. 1.34 (90-degrees phase space rotated: switch x and p). In Fig. 2.25 we can see the odd-photon nature of a minus cat state.

2.5 Creation of a Superposition of Photon-Number States

We can create a superposition of photon-number states with heralding. We will describe that in this section. The simplest example is a superposition of a vacuum $|0\rangle$ and a single-photon state $|1\rangle$.

Figure 2.26 shows the schematic. A photon pair A,B is created by parametric downconversion. As mentioned in Sect. 2.3, it is just a vacuum most of the time and very rarely a photon pair. Photon A enters a beam splitter with transmissivity $T \approx 1$ and the output is combined with a very weak coherent beam, in which the

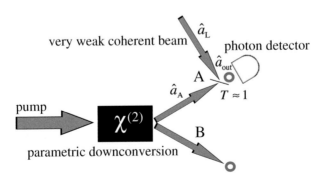

Fig. 2.26 Creation of a superposition of a vacuum $|0\rangle$ and a single-photon state $|1\rangle$. A photon pair A, B is created by parametric downconversion. Photon A enters a beam splitter with transmissivity $T \approx 1$ and the output is combined with a very weak coherent beam, in which the average photon number is much less than one. When we detect a photon at the photon detector, we can get a superposition of a vacuum $|0\rangle$ and a single-photon state $|1\rangle$ in beam B

average photon number is much less than one. When we detect a photon at the photon detector, we can get superposition of a vacuum $|0\rangle$ and a single-photon state $|1\rangle$ in beam B. Note that we assume that we can neglect the probability of simultaneous existence of photons in both beam A and the very weak coherent beam.

We try to make a naive explanation of the mechanism for creation of a superposition of a vacuum $|0\rangle$ and a single-photon state $|1\rangle$ here. When we detect a photon, there are two possibilities. One is that the photon was photon A created by parametric downconversion. Another possibility is that the photon came from the very weak coherent beam. Fortunately or unfortunately the photon detector cannot distinguish the two possibilities, and that is why we can create the superposition. This explanation is really naive and we have to consider entanglement explained in Sect. 1.12 to make the precise explanation. From now on we will deal with entanglement or a two-mode squeezed vacuum (Eq. (1.210)) and try to explain the mechanism for creation of a superposition of a vacuum $|0\rangle$ and a single-photon state $|1\rangle$.

The word "parametric downconversion" is used when the pump in Fig. 2.26 is very weak. In the general case for the pump it becomes a two-mode squeezed vacuum (Eq. (1.210)) explained in Sect. 1.12. We reintroduce Eq. (1.210) here:

$$|\text{EPR}*\rangle = \sqrt{1-q^2} \sum_{n=0}^{\infty} q^n |n\rangle_A \otimes |n\rangle_B. \tag{2.24}$$

Moreover, since a beam splitter matrix with transmissivity T is

$$\begin{pmatrix} \sqrt{T} & \sqrt{1-T} \\ -\sqrt{1-T} & \sqrt{T} \end{pmatrix}, \tag{2.25}$$

the input-output relation of the beam splitter should satisfy

$$\begin{pmatrix} \hat{a}_{\text{out}} \\ \hat{a}_E \end{pmatrix} = \begin{pmatrix} \sqrt{T} & \sqrt{1-T} \\ -\sqrt{1-T} & \sqrt{T} \end{pmatrix} \begin{pmatrix} \hat{a}_A \\ \hat{a}_L \end{pmatrix}, \tag{2.26}$$

where $\hat{a}_A$ represents beam A, $\hat{a}_L$ represents the very weak coherent beam, $\hat{a}_{\text{out}}$ represents one of the output beams from the beam splitter, which is going to the photon detector, and $\hat{a}_E$ represents the other output beam. By using above equations we get

$$\hat{a}_{\text{out}} = \sqrt{T}\hat{a}_A + \sqrt{1-T}\hat{a}_L. \tag{2.27}$$

Here we assume that the very weak coherent beam is in a coherent state $|\beta\rangle$. In the limit of $T \to 1$ and when we assume $|\beta| \to \infty$, $\sqrt{1-T}\beta \to \alpha$, and $|\alpha| \gg 1$, we can treat $\sqrt{1-T}\hat{a}_L$ as α. It is because we can neglect the quantum fluctuations when $|\beta| \gg 1$. So in the limit of $T \to 1$ and $|\beta| \to \infty$, we get

$$\hat{a}_{\text{out}} = \hat{a}_A + \alpha. \tag{2.28}$$

It is obvious that we can make a displacement operation $\hat{D}(\alpha)$ on mode $\hat{a}_A$. We can check it by Eq. (2.7), where we calculate displacement operation in the Heisenberg picture, $\hat{D}^\dagger(\alpha)\hat{a}\hat{D}(\alpha)$.

From above considerations we can see that we are making a displacement operation on one of the two-mode squeezed vacuum in the setup of Fig. 2.26. So the process can be written as

$$
\hat{D}_A(\alpha)\sqrt{1-q^2}\sum_{n=0}^{\infty}q^n|n\rangle_A \otimes |n\rangle_B
$$
$$
= \sqrt{1-q^2}\sum_{n=0}^{\infty}q^n \hat{D}_A(\alpha)|n\rangle_A \otimes |n\rangle_B
$$
$$
= \sqrt{1-q^2}\sum_{n=0}^{\infty}q^n \sum_{m=0}^{\infty}\frac{(\alpha\hat{a}^\dagger - \alpha^*\hat{a})^m}{m!}|n\rangle_A \otimes |n\rangle_B. \qquad (2.29)
$$

When the pump is very weak ($q \ll 1$) and the displacement is very small ($|\alpha| \ll 1$), we get

$$
\sqrt{1-q^2}\sum_{n=0}^{\infty}q^n \sum_{m=0}^{\infty}\frac{(\alpha\hat{a}^\dagger - \alpha^*\hat{a})^m}{m!}|n\rangle_A \otimes |n\rangle_B
$$
$$
\approx (1 + \alpha\hat{a}^\dagger - \alpha^*\hat{a})|0\rangle_A \otimes |0\rangle_B + (1 + \alpha\hat{a}^\dagger - \alpha^*\hat{a})q|1\rangle_A \otimes |1\rangle_B
$$
$$
= \left(|0\rangle_A + \alpha|1\rangle_A\right) \otimes |0\rangle_B + q\left(|1\rangle_A + \alpha\sqrt{2}|2\rangle_A - \alpha^*|0\rangle_A\right) \otimes |1\rangle_B
$$
$$
= |0\rangle_A|0\rangle_B + \alpha|1\rangle_A|0\rangle_B + q|1\rangle_A|1\rangle_B + \sqrt{2}q\alpha|2\rangle_A|1\rangle_B - q\alpha^*|0\rangle_A|1\rangle_B
$$
$$
\approx |0\rangle_A|0\rangle_B + |1\rangle_A \otimes \left(\alpha|0\rangle_B + q|1\rangle_B\right), \qquad (2.30)
$$

where we neglected the terms of $q\alpha$ ($q\alpha^*$) for the last nearly-equality. Therefore when a photon is detected in beam A, beam B becomes[2]

$$
\alpha|0\rangle_B + q|1\rangle_B. \qquad (2.31)
$$

We can thus create a superposition of a vacuum $|0\rangle$ and a single-photon state $|1\rangle$. Here we can say that the displacement operation on beam A was transferred to beam B because of entanglement between beams A and B. Moreover, we can get an arbitrary superposition of a vacuum $|0\rangle$ and a single-photon state $|1\rangle$ by tuning the pump power q and the displacement α. Note that when we set $\alpha = 0$ the output (Eq. (2.31)) becomes $q|1\rangle_B$. We can see that it is heralding creation of a single photon with parametric downconversion in this case.

Figure 2.27 shows an example of phase dependence of amplitude of an experimental superposition of a vacuum $|0\rangle$ and a single-photon state $|1\rangle$ (one cycle). This figure agrees well with Figs. 1.22 and 1.23, which are theoretically calculated superposi-

---

[2]It is not normalized. We should only use the ratio between α and q.

Fig. 2.27 An example of phase dependence of amplitude of an experimental superposition of a vacuum $|0\rangle$ and a single-photon state $|1\rangle$ (one cycle). Note that the method of getting this figure is similar to Fig. 2.26 but not exactly the same

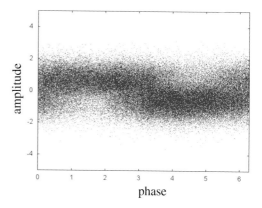

tion states of a vacuum $|0\rangle$ and a single-photon state $|1\rangle$ $((|0\rangle + |1\rangle)/\sqrt{2})$. Moreover, various types of the superposition is examined as shown in Fig. 2.28, where the states are represented by the Wigner functions. These are cases of $|\alpha| : |q| = 1 : 1$ and they are called "qubits" (quantum bits). The results agree well with Fig. 1.44, which is a theoretical calculation. Of course, we can create other states than qubits with just tuning the pump power q and the displacement α.

By using a similar methodology we can create a superposition of many photon-number states. Figure 2.29 shows how to create an arbitrary superposition up to a

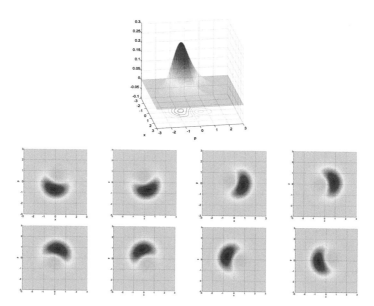

Fig. 2.28 Examples of the Wigner functions of an experimental superposition of a vacuum and a single-photon state $\alpha|0\rangle + q|1\rangle$ ($|\alpha| : |q| = 1 : 1$). It is called (single-rail) qubit

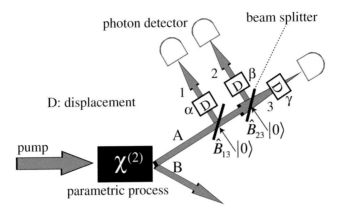

Fig. 2.29 Heralded creation of an arbitrary superposition up to a three-photon state $|3\rangle$. Two-mode squeezed vacuum (beams A, B) is created by a parametric process. Beam A is divided into three beams. Each beam has displacement operation realized by a beam splitter with $T \approx 1$ and a very weak coherent beam as shown in Fig. 2.26. When we have triple coincidence of three photon detector clicks, we can get superposition up to a three-photon state $|3\rangle$

three-photon state $|3\rangle$. In this scheme we use a methodology similar to Fig. 2.26, where we use entanglement and the displacement operation. The "D" in Fig. 2.29 corresponds to the displacement operation and can be realized by a beam splitter with $T \approx 1$ and a very weak coherent beam as shown in Fig. 2.26.

We will explain the mechanism of the heralded creation of an arbitrary superposition up to a three-photon state $|3\rangle$ with Fig. 2.29. We label the three output beams from the two-beam-splitter network as 1, 2, and 3 and describe the two beam-splitter operators as $\hat{B}_{13}$ and $\hat{B}_{23}$ as shown in Fig. 2.29. We set transmissivity of the beam splitters to $T_{13} = 2/3$ and $T_{23} = 1/2$. So the beam splitter transformation becomes

$$\hat{B}_{13}^\dagger \begin{pmatrix} \hat{a}_1 \\ \hat{a}_3 \end{pmatrix} \hat{B}_{13} = \begin{pmatrix} \sqrt{\frac{2}{3}} & \sqrt{\frac{1}{3}} \\ -\sqrt{\frac{1}{3}} & \sqrt{\frac{2}{3}} \end{pmatrix} \begin{pmatrix} \hat{a}_1 \\ \hat{a}_3 \end{pmatrix} \tag{2.32}$$

$$\hat{B}_{23}^\dagger \begin{pmatrix} \hat{a}_2 \\ \hat{a}_3 \end{pmatrix} \hat{B}_{23} = \begin{pmatrix} \sqrt{\frac{1}{2}} & \sqrt{\frac{1}{2}} \\ -\sqrt{\frac{1}{2}} & \sqrt{\frac{1}{2}} \end{pmatrix} \begin{pmatrix} \hat{a}_2 \\ \hat{a}_3 \end{pmatrix}. \tag{2.33}$$

When beam A is in a photon-number state $|n\rangle$, the input state to the two-beam-splitter network can be described as $|0\rangle_1 \otimes |0\rangle_2 \otimes |n\rangle_3 \equiv |0, 0, n\rangle$, because there is no input beam from the other ports of the two-beam-splitter network than beam A as shown in Fig. 2.29.

By using the equations above we can describe the state just before the three photon detectors in Fig. 2.29 as

$$\hat{D}_1(\alpha)\,\hat{D}_2(\beta)\,\hat{D}_3(\gamma)\,\hat{B}_{23}\,\hat{B}_{13}\sqrt{1-q^2}\sum_{n=0}^{\infty}q^n|0,0,n\rangle\otimes|n\rangle_{\mathrm{B}}.\qquad(2.34)$$

To calculate this, we first calculate the following equations by using Eq. (1.193).

$$\hat{B}_{13}|0,1\rangle = \sum_{k_2=0}^{1}\binom{1}{k_2}\left(\sqrt{\frac{1}{3}}\right)^{k_2}\left(\sqrt{\frac{2}{3}}\right)^{1-k_2}\sqrt{k_2!(1-k_2)!}|k_2,1-k_2\rangle$$

$$= \sqrt{\frac{2}{3}}|0,1\rangle + \sqrt{\frac{1}{3}}|1,0\rangle.\qquad(2.35)$$

$$\hat{B}_{13}|0,2\rangle = \frac{1}{\sqrt{2!}}\sum_{k_2=0}^{2}\binom{2}{k_2}\left(\sqrt{\frac{1}{3}}\right)^{k_2}\left(\sqrt{\frac{2}{3}}\right)^{2-k_2}\sqrt{k_2!(2-k_2)!}|k_2,2-k_2\rangle$$

$$= \frac{2}{3}|0,2\rangle + \frac{2}{3}|1,1\rangle + \frac{1}{3}|2,0\rangle.\qquad(2.36)$$

$$\hat{B}_{13}|0,3\rangle = \frac{1}{\sqrt{3!}}\sum_{k_2=0}^{3}\binom{3}{k_2}\left(\sqrt{\frac{1}{3}}\right)^{k_2}\left(\sqrt{\frac{2}{3}}\right)^{3-k_2}\sqrt{k_2!(3-k_2)!}|k_2,3-k_2\rangle$$

$$= \frac{2\sqrt{2}}{3\sqrt{3}}|0,3\rangle + \frac{2}{3}|1,2\rangle + \frac{\sqrt{2}}{3}|2,1\rangle + \frac{1}{3\sqrt{3}}|3,0\rangle.\qquad(2.37)$$

$$\hat{B}_{23}|0,1\rangle = \sum_{k_2=0}^{1}\binom{1}{k_2}\left(\sqrt{\frac{1}{2}}\right)^{k_2}\left(\sqrt{\frac{1}{2}}\right)^{1-k_2}\sqrt{k_2!(1-k_2)!}|k_2,1-k_2\rangle$$

$$= \sqrt{\frac{1}{2}}|0,1\rangle + \sqrt{\frac{1}{2}}|1,0\rangle.\qquad(2.38)$$

$$\hat{B}_{23}|0,2\rangle = \frac{1}{\sqrt{2!}}\sum_{k_2=0}^{2}\binom{2}{k_2}\left(\sqrt{\frac{1}{2}}\right)^{k_2}\left(\sqrt{\frac{1}{2}}\right)^{2-k_2}\sqrt{k_2!(2-k_2)!}|k_2,2-k_2\rangle$$

$$= \frac{1}{2}|0,2\rangle + \frac{1}{\sqrt{2}}|1,1\rangle + \frac{1}{2}|2,0\rangle.\qquad(2.39)$$

$$\hat{B}_{23}|0,3\rangle = \frac{1}{\sqrt{3!}}\sum_{k_2=0}^{3}\binom{3}{k_2}\left(\sqrt{\frac{1}{2}}\right)^{k_2}\left(\sqrt{\frac{1}{2}}\right)^{3-k_2}\sqrt{k_2!(3-k_2)!}|k_2,3-k_2\rangle$$

$$= \frac{1}{2\sqrt{2}}|0,3\rangle + \frac{\sqrt{3}}{2\sqrt{2}}|1,2\rangle + \frac{\sqrt{3}}{2\sqrt{2}}|2,1\rangle + \frac{1}{2\sqrt{2}}|3,0\rangle.\qquad(2.40)$$

By using above equations we get

$$\hat{B}_{23}\hat{B}_{13}\sum_{n=0}^{\infty}q^{n}|0, 0, n\rangle \otimes |n\rangle_{B}$$

$$= |0, 0, 0\rangle \otimes |0\rangle_{B}$$

$$+ q\left[\sqrt{\frac{1}{3}}|0, 0, 1\rangle + \sqrt{\frac{1}{3}}|0, 1, 0\rangle + \sqrt{\frac{1}{3}}|1, 0, 0\rangle\right] \otimes |1\rangle_{B}$$

$$+ q^{2}\left[\frac{1}{3}|0, 0, 2\rangle + \frac{\sqrt{2}}{3}|0, 1, 1\rangle + \frac{1}{3}|0, 0, 2\rangle\right.$$

$$\left. + \frac{\sqrt{2}}{3}|1, 0, 1\rangle + \frac{\sqrt{2}}{3}|1, 1, 0\rangle + \frac{1}{3}|2, 0, 0\rangle\right] \otimes |2\rangle_{B}$$

$$+ q^{3}\left[\frac{1}{3\sqrt{3}}|0, 0, 3\rangle + \frac{1}{3}|0, 1, 2\rangle + \frac{1}{3}|0, 2, 1\rangle\right.$$

$$+ \frac{1}{\sqrt{3}}|1, 0, 2\rangle + \frac{\sqrt{2}}{3}|1, 1, 1\rangle + \frac{1}{3}|1, 2, 0\rangle$$

$$\left. + \frac{1}{3}|2, 0, 1\rangle + \frac{1}{3}|2, 1, 0\rangle + \frac{1}{3\sqrt{3}}|3, 0, 0\rangle\right] \otimes |3\rangle_{B} + \cdots. \qquad (2.41)$$

Now we make three displacement operations on this state. We assume $|\alpha| \ll 1$, $|\beta| \ll 1$, $|\gamma| \ll 1$, and $q \ll 1$ as before, and then we can make the following approximation:

$$\hat{D}(\alpha) \approx 1 + \alpha\hat{a}^{\dagger} - \alpha^{*}\hat{a}. \qquad (2.42)$$

By discarding terms of fourth order and higher of α, β, γ, and q and also discarding terms that does not have at least one photon in each beam (1, 2, 3) we get

$$\hat{D}_{1}(\alpha)\hat{D}_{2}(\beta)\hat{D}_{3}(\gamma)\hat{B}_{23}\hat{B}_{13}\sqrt{1-q^{2}}\sum_{n=0}^{\infty}q^{n}|0, 0, n\rangle \otimes |n\rangle_{B}$$

$$\rightarrow \alpha\beta\gamma|1, 1, 1\rangle \otimes |0\rangle_{B}$$

$$+ q\left[\sqrt{\frac{1}{3}}\alpha\beta|1, 1, 1\rangle + \sqrt{\frac{1}{3}}\alpha\gamma|1, 1, 1\rangle + \sqrt{\frac{1}{3}}\beta\gamma|1, 1, 1\rangle\right] \otimes |1\rangle_{B}$$

$$+ q^{2}\left[\frac{\sqrt{2}}{3}\alpha|1, 1, 1\rangle + \frac{\sqrt{2}}{3}\beta|1, 1, 1\rangle + \frac{\sqrt{2}}{3}\gamma|1, 1, 1\rangle\right] \otimes |2\rangle_{B}$$

$$+ q^{3}\left[\frac{\sqrt{2}}{3}|1, 1, 1\rangle\right] \otimes |3\rangle_{B}$$

$$= |1, 1, 1\rangle \otimes \left[\alpha\beta\gamma|0\rangle_{B} + q\sqrt{\frac{1}{3}}(\alpha\beta + \alpha\gamma + \beta\gamma)|1\rangle_{B}\right.$$

$$\left. + q^{2}\frac{\sqrt{2}}{3}(\alpha + \beta + \gamma)|2\rangle_{B} + q^{3}\frac{\sqrt{2}}{3}|3\rangle_{B}\right]. \qquad (2.43)$$

Therefore, when we have a triple coincidence count of the three photon detectors, the state of beam B becomes

$$\alpha\beta\gamma|0\rangle_B + q\sqrt{\frac{1}{3}}(\alpha\beta + \alpha\gamma + \beta\gamma)|1\rangle_B + q^2\frac{\sqrt{2}}{3}(\alpha + \beta + \gamma)|2\rangle_B + q^3\frac{\sqrt{2}}{3}|3\rangle_B.$$
(2.44)

Here the coefficients originally appearing in beam A appear in beam B after the triple coincidence count. Of course, it is caused by entanglement between beams A and B. Note that the state of Eq. (2.44) is not normalized. So only the ratio between the coefficients has meaning.

We describe the state $|\psi\rangle$ that we want to create as

$$|\psi\rangle = \sum_{n=0}^{3} c_n|n\rangle.$$
(2.45)

To create arbitrary states we have to control c_0, c_1, c_2, and c_3 with the experimental parameters α, β, γ, and q. We can see that it is possible because of the following equations, which can be obtained from the comparison between Eqs. (2.44) and (2.45):

$$\alpha\beta\gamma = c_0,$$
(2.46)

$$q\sqrt{\frac{1}{3}}(\alpha\beta + \beta\gamma + \gamma\alpha) = c_1,$$
(2.47)

$$q^2\frac{\sqrt{2}}{3}(\alpha + \beta + \gamma) = c_2,$$
(2.48)

$$q^3\frac{\sqrt{2}}{3} = c_3.$$
(2.49)

By using the above equation we get

$$q = \left(\frac{3}{\sqrt{2}}c_3\right)^{\frac{1}{3}}.$$
(2.50)

We can get the other experimental parameters by solving the following third order equation:

$$(x - \alpha)(x - \beta)(x - \gamma) = 0.$$
(2.51)

It can be rearranged as follows:

$$x^3 - (\alpha + \beta + \gamma)x^2 + (\alpha\beta + \beta\gamma + \gamma\alpha)x - \alpha\beta\gamma = 0.$$
(2.52)

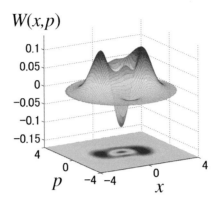

$W(x,p)$

Fig. 2.30 Wigner function of an experimental superposition of a single-photon state $|1\rangle$ and a three-photon state $|3\rangle$ created by the scheme in Fig. 2.29 [8]. It corresponds to a minus cat state with $\alpha = \sqrt{2}$, $N_{\alpha-}(|\alpha\rangle - |-\alpha\rangle)$. It should have at least three negative parts in the Wigner function and we can see them in this figure. Note that we can see only one negative part in Fig. 1.34, where $\alpha = 1$

So we can get α, β, and γ from q, which we already got in Eq. (2.50), by using the following third order equation:

$$x^3 - \frac{3c_2}{\sqrt{2}q^2}x^2 + \frac{\sqrt{3}c_1}{q}x - c_0 = 0. \tag{2.53}$$

Therefore, we can create an arbitrary superposition up to a three-photon state $|3\rangle$ by tuning the pump power and the three displacement parameters. In principle we can extend this methodology up to an arbitrary photon-number state. However, the coincidence rate becomes very small when the photon number increases. So the practical limit is a four-photon state.

Figures 2.30 and 2.31 show examples of Wigner functions of experimental superposition states up to a three-photon state created by the scheme of Fig. 2.29. Figure 2.30 shows Wigner function of an experimental superposition of a single-photon state $|1\rangle$ and a three-photon state $|3\rangle$. As mentioned in Sects. 1.7 and 2.4, a minus cat state $N_{\alpha-}(|\alpha\rangle - |-\alpha\rangle)$ is a superposition of odd photon-number states, and can be created by using single-photon subtraction from a squeezed vacuum. However, since the portions of a two-photon state $|2\rangle$ and a four-photon state $|4\rangle$ in a squeezed vacuum cannot be changed independently, it is impossible to get a minus cat state with $|\alpha| > 1$ and high fidelity by using single-photon subtraction. So we use the scheme in Fig. 2.29 instead of single-photon subtraction to create a "bigger" minus cat state with $|\alpha| > 1$. For example, we can create a "bigger" minus cat state with $\alpha = \sqrt{2}$, which is

$$|1\rangle + \frac{1}{\sqrt{3}}|3\rangle \tag{2.54}$$

without the normalizing factor.

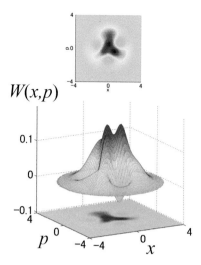

Fig. 2.31 Wigner function of an experimental superposition of a vacuum $|0\rangle$ and a three-photon state $|3\rangle$ created by the scheme of Fig. 2.29 [8]. It might be called "three-headed cat state", because it corresponds to a superposition of three coherent states

Figure 2.30 shows the Wigner function of an experimental superposition of a single-photon state $|1\rangle$ and a three-photon state $|3\rangle$ created by the scheme in Fig. 2.29. It corresponds to a minus cat state with $\alpha = \sqrt{2}$, $N_{\alpha-}(|\alpha\rangle - |-\alpha\rangle)$ as mentioned above. Its Wigner function should at least have three negative parts as we can see in Fig. 2.30. Note that we can see only one negative part in Fig. 1.34, where $\alpha = 1$. Moreover, the fidelity of the Wigner function in Fig. 2.30 is rather high for a minus cat state with $\alpha = \sqrt{2}$.

Another example is shown in Fig. 2.31, which corresponds to a superposition of three coherent states, $|\alpha\rangle + |\alpha e^{i\frac{2\pi}{3}}\rangle + |\alpha e^{-i\frac{2\pi}{3}}\rangle$. We create this state by the scheme in Fig. 2.29. Here we use the equation

$$|\alpha\rangle + |\alpha e^{i\frac{2\pi}{3}}\rangle + |\alpha e^{-i\frac{2\pi}{3}}\rangle \propto |0\rangle + \frac{\alpha^3}{\sqrt{6}}|3\rangle + \cdots, \qquad (2.55)$$

where $\alpha = -1.2i$. We can clearly see three peaks and dips rotated by 120°, which correspond to three coherent states and quantum interference fringes between them, respectively. So it is obvious that three coherent states are superposed in this result.

Before closing this section, we show an experimental three photon state $|3\rangle$, which is also created by the scheme in Fig. 2.29. The example is shown in Fig. 2.32.

$W(x,p)$

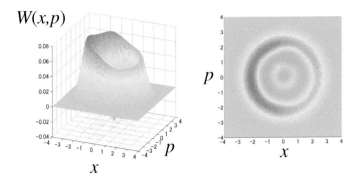

Fig. 2.32 Wigner function of an experimental three-photon state $|3\rangle$ created by the scheme in Fig. 2.29 [8]. The experimental Wigner function agrees with the theoretical one in Fig. 1.39

2.6 Creation of Quantum Entanglement

As mentioned in the previous section, we need to have quantum entanglement between beams A and B for heralding creation of quantum states, and the most important entanglement would be a two-mode squeezed vacuum. It is the nonclassical correlation that causes the photon number of beam B to become n given that the one of beam A is n. Of course, we have other types of correlations, e.g., amplitudes, as shown in Eq. (1.205). In this section we consider a two-mode squeezed vacuum from this point of view.

As shown in Sect. 1.12 the EPR state $|EPR\rangle$ is a maximally entangled state. So if we make an x-measurement on beam A and get the value of x then we know the value of beam B is also x without measuring it ($\hat{x}_A - \hat{x}_A \rightarrow 0$). Similarly, if we make a p-measurement on beam A and get the value of p then we know the value of beam B is $-p$ without measuring it ($\hat{p}_A + \hat{p}_A \rightarrow 0$). Figure 2.33 shows this concept.

So far we have mainly dealt with pure states, where we can use a wavefunction. We extend the story to a general state here. In Sect. 1.12 we explained that the EPR state is the simultaneous eigenstate of $\hat{x}_A - \hat{x}_B$ and $\hat{p}_A + \hat{p}_B$ with zero eigenvalues. It can be described with quantum mechanical expectation values $\langle\,\rangle$ as

$$\langle[\Delta(\hat{x}_A - \hat{x}_B)]^2\rangle + \langle[\Delta(\hat{p}_A + \hat{p}_B)]^2\rangle = 0. \tag{2.56}$$

The point here is that we can use the above condition for all types of states. It is really powerful for experiments.

As mentioned in Sect. 1.12, it is impossible to create a perfect EPR state because we need an infinite amount of energy to do so. However, as also mentioned in Sect. 1.12, we can create an entangled state or a two-mode squeezed vacuum without infinite amount of energy. So we think about loosening the condition of Eq. (2.56).

Fig. 2.33 Concept of
EPR-type correlation
between x_A and x_B and
between p_A and p_B. Each
point corresponds to a
measurement. x is correlated
and p is anti-correlated

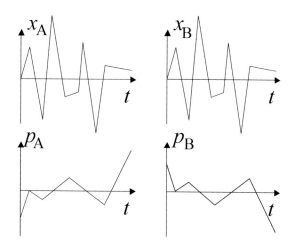

It is well known that beams A and B are entangled if the following condition holds
[9, 10]:

$$\langle [\Delta(\hat{x}_A - \hat{x}_B)]^2 \rangle + \langle [\Delta(\hat{p}_A + \hat{p}_B)]^2 \rangle < 1. \qquad (2.57)$$

Now let's think about the case where beams A and B are not entangled (separable)
and both are in a vacuum state. In this case, since the overall state is $|0\rangle_A \otimes |0\rangle_B$ and
$\langle \Delta\hat{A}^2 \rangle = \langle \psi|\hat{A}^2|\psi\rangle - \langle \psi|\hat{A}|\psi\rangle^2$, we get

$$
\begin{aligned}
\langle [\Delta(\hat{x}_A &- \hat{x}_B)]^2 \rangle + \langle [\Delta(\hat{p}_A + \hat{p}_B)]^2 \rangle \\
&= {}_B\langle 0| \otimes {}_A\langle 0|(\hat{x}_A - \hat{x}_B)^2|0\rangle_A \otimes |0\rangle_B \\
&\quad - \left[{}_B\langle 0| \otimes {}_A\langle 0|(\hat{x}_A - \hat{x}_B)|0\rangle_A \otimes |0\rangle_B \right]^2 \\
&\quad + {}_B\langle 0| \otimes {}_A\langle 0|(\hat{p}_A + \hat{p}_B)^2|0\rangle_A \otimes |0\rangle_B \\
&\quad - \left[{}_B\langle 0| \otimes {}_A\langle 0|(\hat{p}_A + \hat{p}_B)|0\rangle_A \otimes |0\rangle_B \right]^2 \\
&= 1. \qquad (2.58)
\end{aligned}
$$

Here we used Eqs. (1.22), (1.23), (1.24), and (1.25) with $\alpha = 0$. Obviously the
inequality of Eq. (2.57) does not hold.

As an example in which the inequality of Eq. (2.57) holds, we consider a two-
mode squeezed vacuum $|EPR*\rangle_{AB}$. By using the "two-mode squeezing operator"
$\hat{S}'_{AB}(r)$ shown in Sect. 1.12 we will show that a two-mode squeezed vacuum satisfies
the inequality of Eq. (2.57). The "two-mode squeezing operator" $\hat{S}'_{AB}(r)$ is

$$\hat{S}'_{AB}(r) = \hat{B}_{HBS*} \exp\left[\frac{r}{2}\left(\hat{a}_A^{\dagger 2} - \hat{a}_A^2 \right) \right] \exp\left[\frac{r}{2}\left(\hat{a}_B^2 - \hat{a}_B^{\dagger 2} \right) \right]. \qquad (2.59)$$

By using this operator a two-mode squeezed vacuum can be described as $|\text{EPR}*\rangle_{AB} = \hat{S}'_{AB}(r)|0\rangle_A \otimes |0\rangle_B$, which is shown in Eq. (1.210). So we can calculate $\langle [\Delta(\hat{x}_A - \hat{x}_B)]^2 \rangle + \langle [\Delta(\hat{p}_A + \hat{p}_B)]^2 \rangle$ for a two-mode squeezed vacuum as follows:

$$
\begin{aligned}
&\langle [\Delta(\hat{x}_A - \hat{x}_B)]^2 \rangle + \langle [\Delta(\hat{p}_A + \hat{p}_B)]^2 \rangle \\
&= {}_{AB}\langle \text{EPR}*|(\hat{x}_A - \hat{x}_B)^2|\text{EPR}*\rangle_{AB} \\
&\quad - [{}_{AB}\langle \text{EPR}*|(\hat{x}_A - \hat{x}_B)|\text{EPR}*\rangle_{AB}]^2 \\
&\quad + {}_{AB}\langle \text{EPR}*|(\hat{p}_A + \hat{p}_B)^2|\text{EPR}*\rangle_{AB} \\
&\quad - [{}_{AB}\langle \text{EPR}*|(\hat{p}_A + \hat{p}_B)|\text{EPR}*\rangle_{AB}]^2 \\
&= {}_B\langle 0| \otimes {}_A\langle 0|\hat{S}'^{\dagger}_{AB}(r)(\hat{x}_A - \hat{x}_B)^2 \hat{S}'_{AB}(r)|0\rangle_A \otimes |0\rangle_B \\
&\quad - [{}_B\langle 0| \otimes {}_A\langle 0|\hat{S}'^{\dagger}_{AB}(r)(\hat{x}_A - \hat{x}_B)\hat{S}'_{AB}(r)|0\rangle_A \otimes |0\rangle_B]^2 \\
&\quad + {}_B\langle 0| \otimes {}_A\langle 0|\hat{S}'^{\dagger}_{AB}(r)(\hat{p}_A + \hat{p}_B)^2 \hat{S}'_{AB}(r)|0\rangle_A \otimes |0\rangle_B \\
&\quad - [{}_B\langle 0| \otimes {}_A\langle 0|\hat{S}'^{\dagger}_{AB}(r)(\hat{p}_A + \hat{p}_B)\hat{S}'_{AB}(r)|0\rangle_A \otimes |0\rangle_B]^2.
\end{aligned}
\tag{2.60}
$$

Here the following equations holds:

$$
\begin{aligned}
\hat{S}'^{\dagger}_{AB}(r)(\hat{x}_A - \hat{x}_B)^2 \hat{S}'_{AB}(r) &= \hat{S}'^{\dagger}_{AB}(r)(\hat{x}_A - \hat{x}_B)\hat{S}'_{AB}(r)\hat{S}'^{\dagger}_{AB}(r)(\hat{x}_A - \hat{x}_B)\hat{S}'_{AB}(r) \\
&= \left(\frac{e^r \hat{x}_A + e^{-r}\hat{x}_B}{\sqrt{2}} - \frac{e^r \hat{x}_A - e^{-r}\hat{x}_B}{\sqrt{2}} \right)^2 \\
&= 2e^{-2r}\hat{x}_B^2,
\end{aligned}
\tag{2.61}
$$

$$
\begin{aligned}
\hat{S}'^{\dagger}_{AB}(r)(\hat{x}_A - \hat{x}_B)\hat{S}'_{AB}(r) &= \frac{e^r \hat{x}_A + e^{-r}\hat{x}_B}{\sqrt{2}} - \frac{e^r \hat{x}_A - e^{-r}\hat{x}_B}{\sqrt{2}} \\
&= \sqrt{2}e^{-r}\hat{x}_B,
\end{aligned}
\tag{2.62}
$$

$$
\begin{aligned}
\hat{S}'^{\dagger}_{AB}(r)(\hat{p}_A + \hat{p}_B)^2 \hat{S}'_{AB}(r) &= \hat{S}'^{\dagger}_{AB}(r)(\hat{p}_A + \hat{p}_B)\hat{S}'_{AB}(r)\hat{S}'^{\dagger}_{AB}(r)(\hat{p}_A + \hat{p}_B)\hat{S}'_{AB}(r) \\
&= \left(\frac{e^{-r} \hat{p}_A + e^{r}\hat{p}_B}{\sqrt{2}} + \frac{e^{-r} \hat{p}_A - e^{r}\hat{p}_B}{\sqrt{2}} \right)^2 \\
&= 2e^{-2r}\hat{p}_A^2,
\end{aligned}
\tag{2.63}
$$

and

$$
\begin{aligned}
\hat{S}'^{\dagger}_{AB}(r)(\hat{p}_A + \hat{p}_B)\hat{S}'_{AB}(r) &= \frac{e^{-r} \hat{p}_A + e^{r}\hat{p}_B}{\sqrt{2}} + \frac{e^{-r} \hat{p}_A - e^{r}\hat{p}_B}{\sqrt{2}} \\
&= \sqrt{2}e^{-r}\hat{p}_A.
\end{aligned}
\tag{2.64}
$$

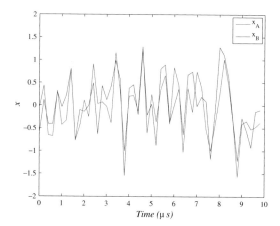

Fig. 2.34 An example of experimental EPR-type correlation or a two-mode squeezed vacuum [11]. Time dependences of x are shown for beams A and B. Each point corresponds to a measurement. We can see strong correlation between x_A and x_B

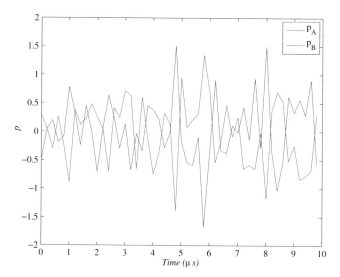

Fig. 2.35 An example of experimental EPR-type correlation or a two-mode squeezed vacuum [11]. Time dependences of p are shown for beams A and B. Each point corresponds to a measurement. We can see strong anti-correlation between p_A and p_B

Note that we use Eqs. (1.157) and (1.209) here.

From these equations we can get

$$\langle [\Delta(\hat{x}_A - \hat{x}_B)]^2 \rangle + \langle [\Delta(\hat{p}_A + \hat{p}_B)]^2 \rangle = e^{-2r} \tag{2.65}$$

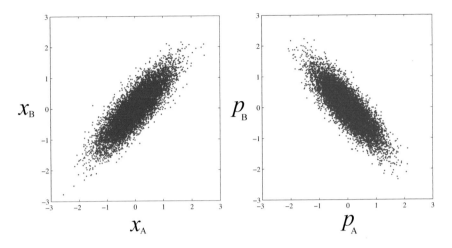

Fig. 2.36 An example of experimental EPR-type correlation or a two-mode squeezed vacuum [11]. We can see strong correlation for x and anti-correlation for p. The degree of (anti-)correlation corresponds to the resource squeezing level e^{-2r}

for a two-mode squeezed vacuum. So if $r > 0$ then beams A and B are entangled. In other words, when the inequality of Eq. (2.57) holds for a two-mode squeezed vacuum $|\text{EPR}*\rangle_{AB}$, we should see the nonclassical correlation of x ($\hat{x}_A - \hat{x}_A \to 0$) and anti-correlation of p ($\hat{p}_A + \hat{p}_A \to 0$) between beams A and B. Although, they are of course not perfect correlations. They become perfect in the limit of $r \to \infty$, where r is the squeezing parameter. Again, r cannot reach infinity, because it needs infinite amount of energy. However, we can increase the r to be high enough. As mentioned in Sect. 2.2, the world record of squeezing level at the moment is $-13\,\text{dB}$. That means we can get $r = 1.5$. Although $r = 1.5$ seems to be very small, the inequality is scaled by e^{-2r} and it is already very small. In that sense even when $r = 1.5$, we can say it is big enough.

We show a last example of experimental EPR-type correlation or a two-mode squeezed vacuum here. The time dependences of x are shown for beams A and B in Fig. 2.34. We can see a strong correlation between x_A and x_B. The time dependences of p are shown for beams A and B in Fig. 2.35. We can see a strong anti-correlation between p_A and p_B. Figure 2.36 shows the correlation plots. So we can see a strong correlation for x and a strong anti-correlation for p. The degree of (anti-)correlation corresponds to the resource squeezing level e^{-2r}. In this experiment we use $-4\,\text{dB}$ of squeezing which corresponds to $r = 0.46$. Even with $r = 0.46$ we can get a reasonably strong EPR-type of correlation. Note that the name "two-mode squeezing" originates from the shapes of the correlation plots of Fig. 2.36.

References

1. R.E. Slusher et al., Phys. Rev. Lett. **55**, 2409 (1985)
2. L.-A. Wu et al., Phys. Rev. Lett. **57**, 2520 (1986)
3. Y. Takeno et al., Opt. Express **15**, 4321 (2007)
4. E. Polzik et al., Appl. Phys. B **55**, 279 (1992)
5. S. Suzuki et al., Appl. Phys. Lett. **89**, 061116 (2006)
6. H. Vahlbruch et al., Phys. Rev. Lett **100**, 033602 (2008)
7. T. Eberle et al., Phys. Rev. Lett. **104**, 251102 (2010)
8. M. Yukawa et al., Opt. Express **21**, 5529 (2013)
9. L.-M. Duan et al., Phys. Rev. Lett. **84**, 2722 (2000)
10. R. Simon, Phys. Rev. Lett. **84**, 2726 (2000)
11. N. Takei et al., Phys. Rev. A **74**, 060101(R) (2006)

© The Author(s) 2015
A. Furusawa, *Quantum States of Light*, SpringerBriefs
in Mathematical Physics, DOI 10.1007/978-4-431-55960-3

Index

© The Author(s) 2015
A. Furusawa, *Quantum States of Light*, SpringerBriefs
in Mathematical Physics, DOI 10.1007/978-4-431-55960-3

Printed in the United States
By Bookmasters